Contents

Probability And Statistics Exam File

Thomas Ward, University of Louisville, Editor; Nazeer Ahmed, University of Louisville; James Bannerman, Southern Technical Institute; Clarence de Silva, Carnegie-Mellon University; Robert Fopma, University of Cincinnati; Gary Harlow, Lehigh University; Kenneth Kerri, California State University, Sacramento; Harold Klee, University of Central Florida; William Lesso, The University of Texas at Austin; John Lovett, The University of Alabama in Huntsville; Hani Mahmassani, The University of Texas at Austin; G. Allen Pugh, Indiana University-Purdue University at Fort Wayne; M. Wayne Parker, Mississippi State University; Charles Proctor, University of Alaska; Jose Sepulveda, University of Central Florida; William Smyer, Mississippi State University

ENGINEERING PRESS, INC. SAN JOSE, CALIFORNIA 95103-0001

Donald G. Newnan, Ph.D.
Exam File Series Editor

Printed in the United States of America

Library of Congress Cataloging in Publication Data
Main entry under title:

Probability and statistics exam file.

(Exam file series)
1. Probabilities--Problems, exercises, etc.
2. Mathematical statistics--Problems, exercises, etc
3. Statistics--Problems, exercises, etc, I. Ward, Thomas L. (Thomas Leon), 1930- . II. Ahmed, Nazeer, 1935- . III. Series.
QA273.25.P76 1985 519.2'076 84-24707
ISBN 0-910554-45-5

5 4 3 2 1

Engineering Press, Inc. P.O. Box 1 San Jose, California 95103-0001

Foreword

Copies of past examinations, particularly those with solutions, are a highly sought after commodity at most schools. In some institutions, copies of final examinations are placed on reserve in the library. As a student, you can benefit substantially from the study of such examinations. First, they are ordinarily drafted in a style that differs from the homework problems in textbooks. Homework problems are written to aid the student in comprehending the material upon which they are based and are intended to be worked in a relatively relaxed atmosphere. Examination problems, on the other hand, are intended to measure comprehension. Since a large number of concepts must be presented in a short time, exam questions may be terse, and are expected to be worked under more stress than homework assignments. Working fast and accurately under stress is a skill that few of us are born with. It must be learned. *Probability and Statistics Exam File* will help you practice your test-taking skills as well as helping with the mastery of the material itself.

The solutions to the problems in the *Exam File* also differ from the solutions to examples in your textbook. Textbook examples are inserted in the text to illustrate immediately proceding concepts and are usually quite detailed. The best and most efficient exam solutions are only sufficiently detailed to demonstrate to your instructor that you mastered the material. One should "read" textbook examples for comprehension. On the other hand, the best way to use past exams is to actually "take" them. Cover the *Exam File* solution and attempt to write out your own solution. Then compare your solutions with those proposed by the instructor. Mark any that you have difficulty with, and return to them the next day.

You may believe that you can "calibrate" an instructor by studying examples of his or her past exams. But many faculty members constantly write new examination questions to probe particular areas that have been developed in the most recent term. We have attempted to deal with this variability by obtaining exam questions from professors from around the country. These professors all teach undergraduate courses in probability and statistics and rely on less than a dozen textbooks. We have ask them to submit questions that they have used and "tested" on actual examinations. For each question there is also a solution prepared by the same instructor that wrote the question. This procedure not only increases the usefulness of the questions, but serves to minimize errors. Inevitably a few errors will creep in. If you find one, drop us a line at Engineering Press. We guarantee a reply!

Thomas L. Ward
Editor

1 DESCRIPTIVE STATISTICS

CUMULATIVE FREQUENCY DISTRIBUTIONS

1-1

The following sample data for completing the framing of a standard tract house are incomplete:

Time (days)	Number of Houses	Relative Frequency	Cumulative Frequency
5-under 6	--	--	--
6-under 7	19	--	27
7-under 8	--	.35	--
8-under 9	15	.15	--
9-under 10	--	--	91
10-under 11	--	--	--
TOTALS	100	1.00	

Determine the missing values and fill in the above table.

****

Cell #	Time	f_i Frequency	P_i Relative Frequency	F_i Cumulative Frequency
1	5-under 6	8 ①	.08 ②	8 ③
2	6-under 7	19	.19 ④	27
3	7-under 8	35 ⑤	.35	62 ⑥
4	8-under 9	15	.15	77 ⑦
5	9-under 10	14 ⑧	.14 ⑨	91
6	10-under 11	9 ⑩	.09 ⑪	100 ⑫

Note that the "number of houses" in a given cell i is the frequency f_i of that cell.

Explanation

① $f_1 = F_2 - f_2 = 27 - 19 = 8$

② $p_1 = \frac{f_1}{\text{sample size}} = \frac{8}{100} = 0.08$

③ $F_1 = f_1 = 8$

④, ⑨ & ⑪ same logic as ②

⑤ $f_3 = p_3 \times \text{sample size} = 35$

⑥ $F_3 = F_2 + f_3 = 27 + 35 = 62$

⑦ $F_4 = F_3 + f_4 = 62 + 15 = 77$

⑧ $f_5 = F_5 - F_4 = 91 - 77 = 14$

⑩ $f_6 = 100 - F_5 = 100 - 91 = 9$

⑫ $F_6 = 100$ (highest cell)

1-2

The diameters in centimeters of a sample of 300 ball bearings manufactured by one company are shown in the frequency table below.

Diameter, cm	0.435	0.440	0.445	0.450	0.455	0.460	0.465	0.470	0.475
Number of ball bearings	2	13	26	48	79	69	41	17	5

Calculate the mean from the above data.

****

X	f	$X' = 1000X - 435$	$X'' = X'/5$	fX''
0.435	2	0	0	0
0.440	13	5	1	13
0.445	26	10	2	52
0.450	48	15	3	144
0.455	79	20	4	316
0.460	69	25	5	345
0.465	41	30	6	246
0.470	17	35	7	119
0.475	5	40	8	40
	300			1275

$\bar{X}'' = \frac{\Sigma f X''}{\Sigma f} = \frac{1275}{300} = 4.25$ $\qquad \bar{X}' = 5\bar{X}'' = 5(4.25) = 21.25$

$\bar{X}' = 1000\bar{X} - 435$ $\qquad \bar{X} = \frac{\bar{X}' + 435}{1000} = \frac{21.25 + 435}{1000} = 0.456 \text{ cm}$

MEASURES OF POSITION

1-3

The following values represent the cycle times observed during a repetitive factory task, in minutes, in the order recorded:

4.5, 3.2, 3.8, 4.7, 5.1, 3.3, 4.2, 2.8, 3.5, 4.4, 3.7, 2.9, 2.8,
3.2, 3.5, 4.3, 3.0, 4.1, 3.2, 4.0

Starting at the minimum value in the data, set up class intervals of width 0.4 minute and find the cumulative frequency in the fifth interval, the mode, and the median of the data.

****

Array (order) the values as follows:

2.8	2.8	2.9	3.0	3.2	3.2	3.2	3.3	3.5	3.5
3.7	3.8	4.0	4.1	4.2	4.3	4.4	4.5	4.7	5.1

Class Interval	Frequency	Cumulative Frequency
2.8 - 3.2	4	4
3.2 - 3.6	6	10
3.6 - 4.0	2	12
4.0 - 4.4	4	16
4.4 - 4.8	3	19
4.8 - 5.2	1	20

Cumulative frequency in fifth interval = 19

Mode (most frequently occurring value) = 3.2

Median (middle value, or average of middle values, in arrayed data) = average of 3.5 and 3.7

= 3.6

1-4

Weights of 30 students in a high school class are recorded in kilograms:

```
46.2, 45.0, 43.2, 42.9, 41.6, 47.7, 40.0, 42.7, 48.3, 50.0,
45.0, 44.5, 42.1, 46.5, 46.9, 48.0, 44.2, 41.0, 43.0, 44.1,
47.5, 45.8, 44.8, 45.7, 43.2, 46.3, 45.9, 46.8, 47.6, 45.9
```

Chose a suitable class interval for this data and prepare a frequency table having at least 8 classes of data.

(i) What is the mode class?
(ii) Using the mode-class mark as the assumed mean to code the data, calculate an estimate for the mean of the given data.
(iii) Calculate the mean value using all 30 values of data and determine the percentage error due to data grouping.

****

The minimum data value is 40.0 ad the maximum data value is 50.0. Divide this interval into 10 subintervals, each having a "class interval" of 1.0. The frequency table is given below:

Class Interval	Class Mark x_i	Frequency f_i	Deviation $d_i = x_i - x_0$ (coded data)	$f_i d_i$	Cumulative Frequency F_i
40 - 41⁻	40.5	1	−5.0	−5	1
41 - 42⁻	41.5	2	−4.0	−8	3
42 - 43⁻	42.5	3	−3.0	−9	6
43 - 44⁻	43.5	3	−2.0	−6	9
44 - 45⁻	44.5	4	−1.0	−4	13
45 - 46⁻	45.5	6	0	0	19
46 - 47⁻	46.5	5	+1.0	+5	24
47 - 48⁻	47.5	3	+2.0	+6	27
48 - 49⁻	48.5	2	+3.0	+6	29
49 - 50	49.5	1	+4.0	+4	30
Σ		30		−11	

(i) Mode corresponds to the maximum frequency (6 in this problem). Hence

Mode class = 45 – 46⁻

(ii) Mode-class mark = 45.5

Assumed mean x_0 = 45.5

The deviations d_i of x_i from x_0 are known as the coded data. From the 5th column of the frequency table,

Correction value for mean ($\Sigma d_i / \Sigma f_i$) $= -\frac{11}{30} = -0.37$

$\therefore$ Estimated mean $= x_0 + \frac{\Sigma d_i}{\Sigma f_i} = 45.5 - 0.37$

$= 45.13$

(iii) Using the entire set of 30 data values

Mean $= \frac{1}{30}[46.2 + 45.0 + \cdots + 47.6 + 45.9]$

$= 45.08$

Percentage error $= \frac{(45.13 - 45.08)100}{45.08}\%$

$= 0.11\%$

1-5

A water pump manufacturing company sells Model 1 at \$100 per unit, Model 2 at \$125 per unit, Model 3 at \$175 per unit, Model 4 at \$250 per unit and Model 5 at \$600 per unit.

Based on a sample of 1000 units sold, the relative frequency distribution of sales of various models is 0.3, 0.25, 0.22 and 0.13 for models 1, 2, 3 and 4 respectively.

a) After computing the relative frequency of Model 5, find the cumulative frequency distribution of models sold.

b) Compute the average selling price per unit, the median selling price per unit and the corresponding mode. What can you say about the skewness of this distribution?

****

a)

Model	Price	Relative Freq.	Frequency	Cumulative Freq.
1	\$ 100	0.3	300	300
2	125	0.25	250	550
3	175	0.22	220	770
4	250	0.13	130	900
5	600	0.10	100	1000
		1.00	1000	

b) Let x_i denote the selling price of model i, and f_i its frequency.

Average selling price

$$\bar{X} = \frac{\sum_{i=1}^{5} x_i f_i}{1000} = \$192.25$$

median selling price = \$125

mode = \$100 (highest frequency)

This distribution is <u>skewed right</u> (mean is to the right of the mode)

MEASURES OF DISPERSION

1-6

Find $\bar{x}$ and S^2 for

x_i	35	40	45	50	55
f_i	13	11	14	13	12

****

FOR SIMPLICITY, CODE X_i by $u_i = \dfrac{x_i - 45}{5}$

DISTRIBUTION THEN BECOMES:

u_i	−2	−1	0	1	2
f_i	13	11	14	13	12

$\Sigma f_i = 63$

COMPUTE $\Sigma f_i \cdot u_i = 0$ AND $\Sigma f_i u_i^2 = 124$

THEN $\bar{u} = 0/63 = 0$

$$S_u^2 = \frac{\Sigma f_i u_i^2 - \frac{(\Sigma f_i \cdot u_i)^2}{\Sigma f_i}}{\Sigma f_i - 1} = \frac{124 - 0}{62} = 2$$

TO DECODE: $\bar{x} = 5 \cdot \bar{u} + 45 = 45$

$S_x^2 = 5^2 \cdot S_u^2 = 25 \cdot 2 = 50$

1-7

In a comparative study of traffic congestion in U.S. cities, the time needed to drive from the Central Business District (CBD) to the airport in two different cities has been observed for 50 working days during the afternoon rush hour. The respective frequency distributions of the CBD-airport travel time for the two cities are shown hereafter:

	Frequency	
Travel time (in minutes)	City 1	City 2
10.0 to under 15.0	5	5
15.0 to under 20.0	10	15
20.0 to under 25.0	30	20
25.0 to under 30.0	5	10

1) Calculate the mean travel times in each city and the corresponding coefficients of variation. Which city exhibits greater variability in the travel time from the CBD to the airport?

2) Find the cumulative frequency distribution for the sample obtained in city 1, and sketch it graphically. What is the median cell for this sample?

3) Find the relative frequency distribution for city 2.

****

1) Let $\bar{X}_i$ = mean travel time in City i , i = 1, 2
s_i = std. dev. of travel time in City i , i = 1, 2

$$\bar{X}_1 = (12.5 \times 5 + 17.5 \times 10 + 22.5 \times 30 + 27.5 \times 5)/50 = 21.0 \text{ mins.}$$

$$s_1 = \left[\frac{(12.5-21.0)^2 \times 5 + (17.5-21.0)^2 \times 10 + (22.5-21.0)^2 \times 30 + (27.5-21.0)^2 \times 5}{49}\right]^{1/2}$$

$$= 3.945 \text{ mins.}$$

coefficient of variation in City 1 = $\frac{s_1}{\bar{X}_1} \times 100\% = 18.78\%$

$\bar{X}_2 = 21.0$ mins , $s_2 = 4.546$ mins.

$\Rightarrow$ coefficient of variation in City 2 = 21.65 %

City 2 exhibits greater variability than City 1, since it has the same mean but a greater coeff. of variation.

2) cumulative frequency distribution for city 1, F_1

Travel time	Cumulative Frequency
10.0 – 15.0	5
15.0 – 20.0	15
20.0 – 25.0	45
25.0 – 30.0	50

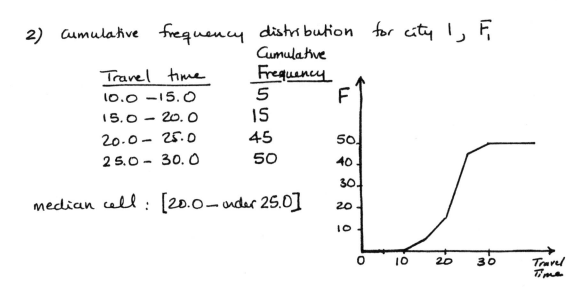

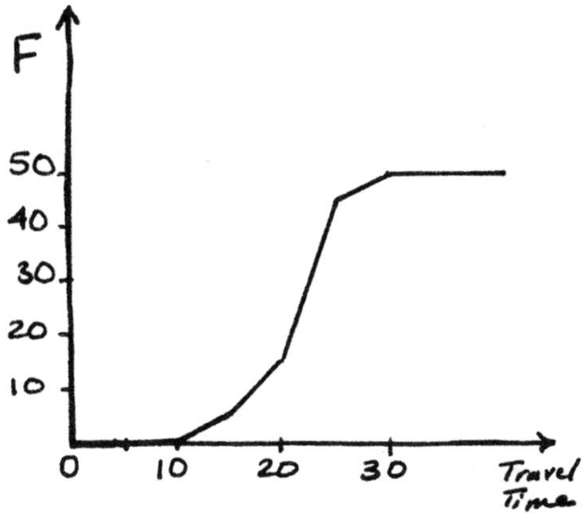

median cell : [20.0 – under 25.0]

3) Relative frequency distribution for city 2;

Travel time	Relative Frequency
10.0 – 15.0	0.10
15.0 – 20.0	0.30
20.0 – 25.0	0.40
25.0 – 30.0	0.20

1-8

Without arranging the data in a frequency distribution, calculate a) $\bar{x}$, b) R, c) M_i, d) S^2, for the following set of weights:

38, 50, 37, 44, 46, 53, 48, 38, 42, 46, 42

****

a) $\bar{x} = \dfrac{\Sigma x}{n} = 44$

b) R(RANGE) = 53 – 37 = 16

c) M_i (MEDIAN) = 44

d) $s^2 = \dfrac{n \Sigma x_i^2 - (\Sigma x_i)^2}{n(n-1)} = 27$

1-9

The diameters in centimeters of a sample of 300 ball bearings manufactured by one company are shown in the frequency table below.

Diameter, cm	0.435	0.440	0.445	0.450	0.455	0.460	0.465	0.470	0.475
Number of ball bearings	2	13	26	48	79	69	41	17	5

Calculate the standard deviation from the above data.

****

X	f	$X' = 1000X - 435$	$X'' = X'/5$	fX''	fX''^2
0.435	2	0	0	0	0
0.440	13	5	1	13	13
0.445	26	10	2	52	104
0.450	48	15	3	144	432
0.455	79	20	4	316	1264
0.460	69	25	5	345	1725
0.465	41	30	6	246	1476
0.470	17	35	7	119	833
0.475	5	40	8	40	320
	300			1275	6167

$$s''^2 = \frac{\Sigma f X''^2 - (\Sigma f X'')^2/\Sigma f}{\Sigma f - 1} = \frac{6167 - (1275)^2/300}{300 - 1}$$

$$= \frac{748.25}{299} = 2.50$$

$$s'' = 1.58$$

$$s = \frac{(1.58)(5)}{1000} = \underline{\underline{0.0079 \text{ cm}}}$$

1-10

A machine is producing metal pieces that are cylindrical in shape. A sample of pieces is taken and the diameters are 1.01, .97, 1.03, 1.04, 0.99, 0.98, 0.99, 1.01, and 1.03 cm. Find the mean, median, range, variance and standard deviation for the sample.

****

$\bar{X} = \dfrac{\sum X_i}{n} = 1.0056$ MEAN

$\tilde{X} = 1.01$ MIDDLE VALUE – MEDIAN

$R = 0.07$ $X_{LARGEST} - X_{SMALLEST}$

$S^2 = \dfrac{n \sum X_i^2 - (\sum X_i)^2}{n(n-1)} = 6.0278$ VARIANCE

$S = \sqrt{6.0278} = 2.455$ STANDARD DEVIATION

med = .97, .98, .99, .99, 1.01, 1.01, 1.03, 1.03, 1.04

mean =

$\frac{81.9459 - (9.05)^2}{9(8)} = \frac{81.9025}{9(8)}$

MEASURES OF SKEWNESS

1-11

The table below gives the number of miles driven per day by a package delivery service in a small city during the month of April:

202	176	124	144	150	175	178	102	166	174
125	182	152	135	174	181	101	115	184	171
209	160	192	173	157	147	183	150	161	168

Using the equations for ungrouped data compute the following statistics:

a) The Mean
b) The Median
c) The Mode
d) The Range
e) The Variance
f) The Standard Deviation
g) The Pearson Coefficient of Skewness

****

a) The mean $\bar{X} = \frac{\Sigma X_i}{n} = \frac{4811}{30} = \underline{160.37}$

b) The median = middle value in an array $= \frac{166+168}{2} = \underline{167}$

c) The mode = most frequent value $\underline{150 \text{ and } 174}$ (bimodal)

d) The range = largest − smallest $= 209 - 101 = \underline{108}$

e) The variance $= s^2 = \frac{\Sigma (X_i - \bar{X})^2}{n-1} = \frac{21516.97}{29} = 741.96$

f) The Standard deviation $= s = \sqrt{s^2} = \sqrt{741.96} = \underline{27.24}$

g) The Pearson Coefficient of Skewness $= Sk = \frac{3(\bar{X} - md)}{s} = \frac{3(160.37 - 167)}{27.24}$

$= \underline{-.73}$

1-12

The following classified data represents the number of miles (round trip) driven per day by 100 workers in a Government office.

miles	frequency
9.5-12.5	2
12.5-15.5	1
15.5-18.5	2
18.5-21.5	1
21.5-24.5	6
24.5-27.5	7
27.5-30.5	11
30.5-33.5	12
33.5-36.5	12
36.5-39.5	17
39.5-42.5	19
42.5-45.5	9
45.5-48.5	0
48.5-51.5	0
51.5-54.5	1

Using the equations for grouped sample data, compute the following statistics:

a) The Mean
b) The Median
c) The Variance
d) The Standard Deviation
e) The Pearson Coefficient of Skewness

****

Class	f_i	X_i	$X_i f_i$	$(X_i-\bar{X})$	$(X_i-\bar{X})^2$	$(X_i-\bar{X})^2 f_i$
9.5-12.5	2	11	22	-23.07	532.22	1064.45
12.5-15.5	1	14	14	-20.07	402.80	402.80
15.5-18.5	2	17	34	-17.07	291.38	582.77
18.5-21.5	1	20	20	-14.07	197.96	197.96
21.5-24.5	6	23	138	-11.07	122.54	735.27
24.5-27.5	7	26	182	-8.07	65.12	455.87
27.5-30.5	11	29	319	-5.07	25.70	282.75
30.5-33.5	12	32	384	-2.07	4.28	51.42
33.5-36.5	12	35	420	.93	.86	10.38
36.5-39.5	17	38	646	3.93	15.44	262.56
39.5-42.4	19	41	779	6.93	48.02	912.47
42.5-45.5	9	44	396	9.93	98.60	887.44
45.5-48.5	0	47	0	12.93	167.18	.00
48.5-51.5	0	50	0	15.93	253.76	.00
51.5-54.5	1	53	53	18.93	358.34	358.34
		SUM	3407			6204.51

Using values from the computation table above we obtain

a) the mean $= \bar{X} = \frac{\Sigma X_i f_i}{n} = \frac{3407}{100} = \underline{34.07}$

b) the median The 50th data point is located in the 33.5 to 36.5 class interval

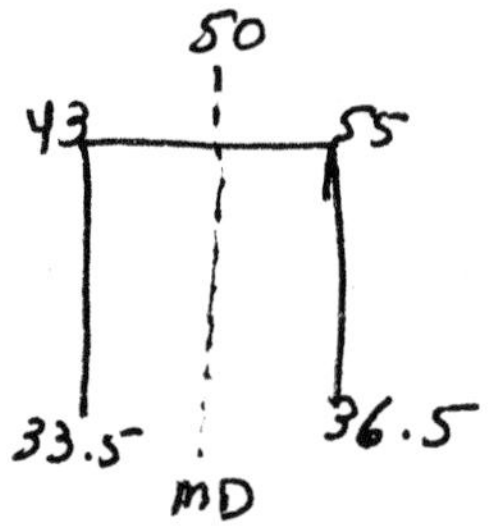

usin ratio and proportion $\frac{md - 33.5}{36.5 - 33.5} = \frac{50 - 43}{55 - 43}$

$md = 33.5 + 3\left(\frac{7}{12}\right) = \underline{35.25}$

c) The variance $s^2 = \frac{\Sigma(X_i - \bar{X})^2 f_i}{n-1} = \frac{6204.51}{99} = \underline{62.67}$

d) the standard deviation $S = \sqrt{s^2} = \sqrt{62.67} = \underline{7.92}$

e) the Pearson Coefficient of Skewness

$S_K = \frac{3(\bar{X} - md)}{s} = \frac{3(34.07 - 35.25)}{7.92} = .45$

The distribution is slightly skewed to the left

2
PROBABILITY

SAMPLE SPACES

2-1

Four democrats are running in the presidential primaries. Suppose that they could be ranked 1, 2, 3, and 4 in the order of their competence, 1 denoting the best. Two of these candidates end up in the democratic ticket (president and vice president) that year. Determine the probability that the ticket

(i) carries the two most competent candidates
(ii) does not carry the most competent candidate

Assume that the voters are ignorant of the competence of the candidates and, as a result, each candidate has the same chance of being elected.

****

The possible distinct choices are

(1, 2) (2, 3) (3, 4)
(1, 3) (2, 4)
(1, 4)

These six pairs of elements (simple events) form the "sample space". Since the selection of each pair is assumed purely random, each event has the same probability 1/6.

NOTE: We don't have to identify the president and the vice president in a given ticket (event). Hence the order of the pair of numbers is irrelevant. Even if we considered the order, we would get the same answer except that the sample space would then consist of 12 simple events, each having the probability 1/12.

(i)

Probability of picking the best ticket

$= P[(1,2)] = \underline{1/6}$

(ii)

Probability of selecting the best candidate

$= P[(1,2) \text{ or } (1,3) \text{ or } (1,4)] = 1/6 + 1/6 + 1/6 = 1/2$

$\therefore$ Probability of not selecting the best $= 1 - 1/2 = \underline{1/2}$

COPLEMENTATION

2-2

Error in this

A CLASS CONSISTS OF 60 STUDENTS. OF THE 35 BORN IN TEXAS, 20 ARE UNDER 20 YEARS OF AGE. A TOTAL OF 25 ARE 20 YEARS OF AGE OR OLDER. DRAW THE VENN DIAGRAM AND DETAIL ALL AREAS.

HOW MANY ARE NOT BORN IN TEXAS AND 20 YEARS OLD OR OLDER ?

****

A ----> BORN IN TEXAS

B ----> 20 YEARS OR OLDER

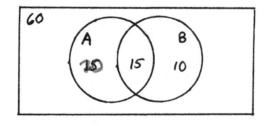

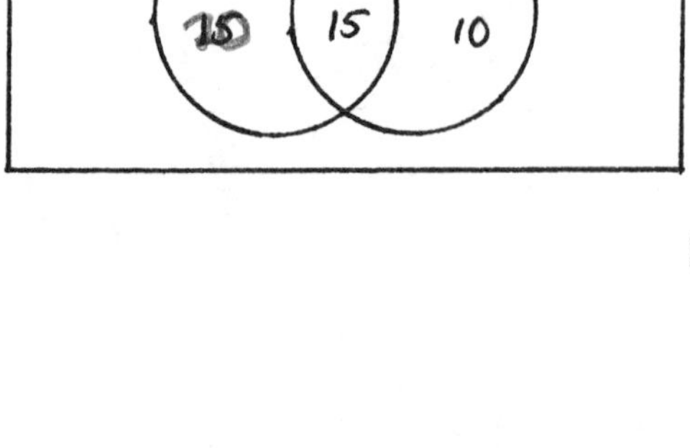

$N(\bar{A} \text{ and } B) = 10$

2-3

An experiment consists of recording the number n of vehicles waiting at a signalized intersection at the instant that the signal indication changes from Red to Green. We define the following events:

$A = \{n > 0\}$
$B = \{1 \leq n \leq 5\}$
$C = \{4 < n \leq 15\}$
$D = \{6 \leq n < 21\}$

a) (i) Are events B, C and D collectively exhaustive? Justify your answer.

(ii) Are events A and B collectively exhaustive? Justify your answer.

b) Specify the elements contained in the following events:

(i) $B \cap C \cap D$

(ii) $(A \cap B) \cup (C \cap D)$

(iii) $A^c \cap D$

****

a) (i) B, C and D are not collectively exhaustive because $\{B \cup C \cup D\}$ does not include the element "0" nor values of $n \geq 21$.

(ii) A and B are not collectively exhaustive because $\{A \cup B\}$ does not include the element "0".

b) (i) $B \cap C \cap D = \{\emptyset\}$

(ii) $(A \cap B) \cup (C \cap D) = \{1 \leq n \leq 15\}$

(iii) $A^c \cap D = \{\emptyset\}$

MUTUAL EXCLUSIVITY

2-4

A random sample showed that 33 students preferred the quarter system over the semester system. If the sample consisted of 50 males and 35 females, then how many favored the semester system? If 23 males preferred the semester system, how many females favored the quarter system?

****

CONSIDER A VENN DIAGRAM:

THE SETS OF THOSE WHO FAVOR ONE SYSTEM OVER THE OTHER ARE MUTUALLY EXCLUSIVE.

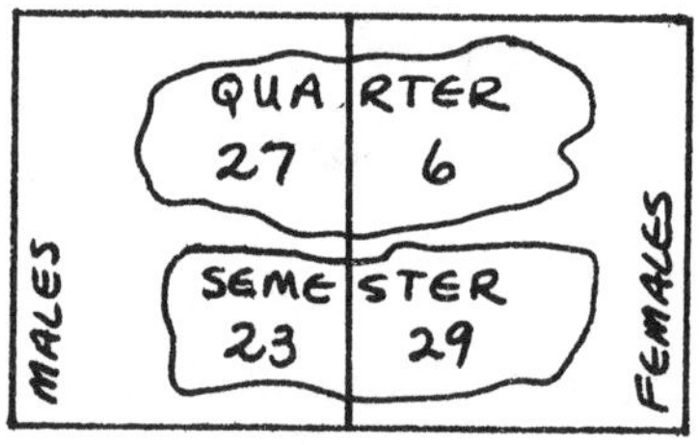

85 TOTAL
− 33 QUARTER
52 SEMESTER OF WHICH 23 ARE MALES
29 ARE FEMALES

35 FEMALES
− 29 FEMALES − SEMESTER
6 FEMALES − QUARTER

2-5

Let N denote the number of fatal accidents in a given year in a particular community.

Let events A = $(N < 20)$

B = $(20 \leq N < 30)$

C = $(25 < N < 30)$

D = $(25 \leq N \leq 45)$

E = $(N \geq 50)$

a) Specify the elements of events $(A \cap C)$, $(B \cap C)$, $(A \cup B)$, and $(B \cap C \cap D)$.

b) Answer the following questions, justifying your answer in each case:

i) Is the set of events A,B,C,D and E collectively exhaustive?

ii) Are events B and D mutually exclusive?

iii) Are events A and D independent?

iv) Without performing any calculations, what can you conclude about the relative probabilities of events C and B?

****

a) $A \cap C = \{\emptyset\}$ $\quad B \cap C = \{25 < N < 30\} = C$

$A \cup B = \{N < 30\}$ $\quad B \cap C \cap D = \{25 < N < 30\} = C$

b) (i) No, because $\{45 < N < 50\}$ is not contained in $\{A \cup B \cup C \cup D \cup E\}$.

(ii) No; $B \cap D = \{25 \leq N < 30\} \neq \{\emptyset\}$

(iii) No, because they are mutually exclusive, and mutually exclusive events <u>cannot</u> be independent.

(iv) Since C is a subset of B, then $p(C) \leq p(B)$.

BASIC SET THEORY

2-6

There are 50 students in a particular engineering class. 30% of the students receive financial aid through the university and 20 members of the class are graduate students. Thirty five of the students own a programmable calculator and three of the five graduate students receiving financial aid own a programmable calculator. Twenty undergraduate students not on financial aid own programmable calculators. Nine graduate students neither receive financial aid nor own a programmable calculator. How many undergraduate students who receive financial aid also own programmable calculators?

****

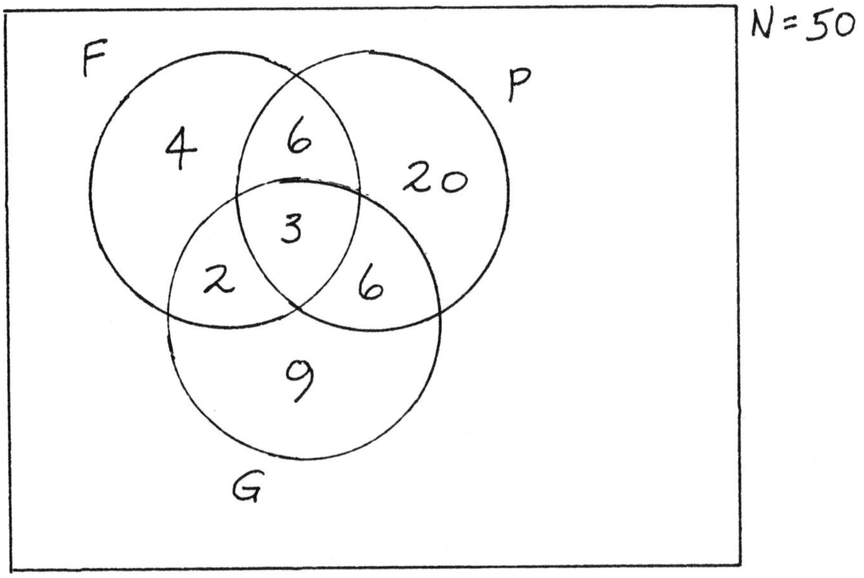

Events: F = receives financial aid
P = owns programmable calculator
G = is a graduate student

From the Venn diagram,

$$N(F \cap P \cap G') = 6$$

2-7

In a class of 74 students, 30 students take Literature, 35 students take Physics, and 40 students take Mathematics. If the corresponding sets are denoted by A, B, and C respectively, the usual notation is n(A) = 30, n(B) = 35, and n(C) = 40. The number of students who take both Literature and Physics is 15. The number of students who take all three subjects is 5. The number of students who take Mathematics but neither Literature nor Physics is 20. The number of students who take Physics and Mathematics but not Literature is 8. Represent these data using a Venn diagram and indicate the number of elements in each disjoint subset.

(i) How many students take Literature but neither Physics nor Mathematics?

(ii) How many students do not take any of the three subjects?

****

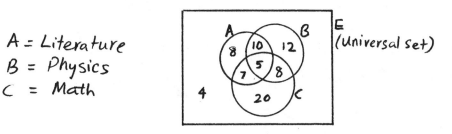

A = Literature
B = Physics
C = Math

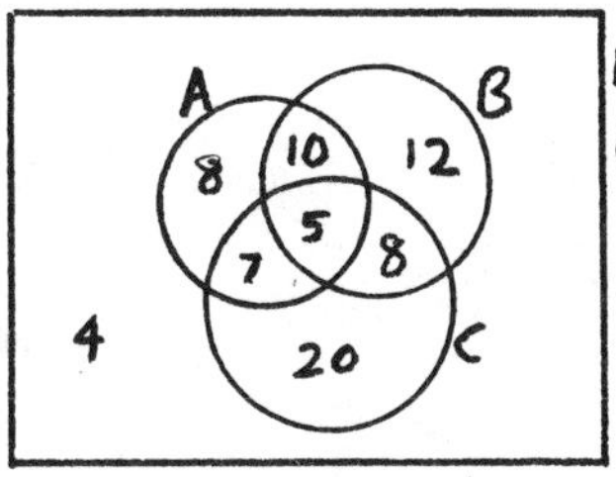

E (Universal set)

(i) Set taking Physics or Math $= B \cup C$

Set not taking either one of these two subjects $= (B \cup C)'$

$\therefore$ The set in question $= A \cap (B \cup C)'$

From the Venn diagram $n(A \cap (B \cup C)') = \underline{\underline{8}}$

(ii) Set taking any of the three subjects $= A \cup B \cup C$

$\therefore$ The set in question $= (A \cup B \cup C)'$

From the Venn diagram $n(A \cup B \cup C)' = \underline{\underline{4}}$

COMBINATORIAL ANALYSIS: PREMUTATIONS

2-8

To successfully route luggage from one airport to another, the destination airport is identified by three letters. For example, Los Angeles is LAX. How many airports could be served by this identification system? State any assumptions.

****

Assume each letter may be repeated (Sampling with replacement)

No. of Airports = (26)(26)(26) = 17,576 Airports

Assume each letter may not be repeated (Sampling without replacement)

No. of Airports = (26)(25)(24) = 15,600 Airports

COMBINATORIAL ANALYSIS: COMBINATIONS

2-9

A product has 5 ways in which it might have a critical defect and 10 ways in which it might have a minor defect. In how many different ways may it have: a) one critical defect and two minor defects, b) three minor defects, c) two critical defects and four minor defects.

****

(a) $\binom{5}{1}\binom{10}{2} = \frac{5!}{4!\,1!} \cdot \frac{10!}{8!\,2!} = 5 \cdot \frac{10 \cdot 9}{1 \cdot 2} = 225$

(b) $\binom{10}{3} = \frac{10!}{7!\,3!} = \frac{10 \cdot 9 \cdot 8}{3 \cdot 2 \cdot 1} = 120$

(c) $\binom{5}{2}\binom{10}{4} = \frac{5!}{3!\,2!} \; \frac{10!}{6!\,4!} = 2100$

2-10

How many different bowling teams of four people each can be formed from a group of 35 bowlers?

****

Since position on the team makes no difference the combination of 35 people taken 4 at a time is appropriate.

$$C_r^n \; \frac{n!}{r!(n-r)!}$$

$$C_4^{35} = \frac{35!}{4!(31!)} = \frac{35 \cdot 34 \cdot 33 \cdot 32 \cdot (31!)}{4 \cdot 3 \cdot 2 \cdot 1\,(31!)} = 52{,}360 \text{ teams}$$

2-11

A Transportation Engineering consulting firm consists of 40 engineers divided among 4 divisions: 15 engineers in Systems Analysis and Planning (SAP), 12 in Traffic Engineering Operations and Control (TEOC), 8 in Geometrics, Pavements and Materials (GPM) and 5 in Computers and Statistics (CS).

a) A team consisting of one engineer from each division is to be formed to undertake a freeway improvement project. How many different teams can be formed for this project?

b) The firm has been asked to perform a feasibility study for a new major urban rapid transit system. A team consisting of 4 SAP, 2 TEOC and 2 CS engineers is needed for this study. How many different teams can be formed for this project?

****

a) # of teams $= \underset{(SAP)}{15} \times \underset{(TEOC)}{12} \times \underset{(GPM)}{8} \times \underset{(CS)}{5} = 7{,}200$

b) # of teams $= \binom{15}{4}\binom{12}{2}\binom{5}{2}$

$$= \frac{15!}{4!\,11!}\;\frac{12!}{2!\,10!}\;\frac{5!}{2!\,3!} = 1365 \times 66 \times 10 = 900{,}900 .$$

2-12

A bin contains 50 bolts, 10 of which are defective. Assume that the bolts are separately identifiable (may be numbered 1 to 50). A robot grabs 5 bolts simultaneously from the bin at random. What is the probability that not more than two bolts picked by the robot are defective?

****

The order in which the five bolts are picked is not relevant in this problem.

Total number of possible "combinations" of five

$$= {}^{50}C_5 = \frac{50!}{5!\,45!}$$

There are 40 good bolts. The number of combinations of five from the good ones

$$= {}^{40}C_5 = \frac{40!}{5!\,35!}$$

If one bolt in the sample of 5 is defective;

The defective one can be picked in ${}^{10}C_1$ ways. For each such selection the remaing good 4 can be picked in ${}^{40}C_4$ ways.

$\therefore$ Total number of ways $= {}^{10}C_1\,{}^{40}C_4$

If two bolts in the sample of 5 are defective;

Total number of ways $= {}^{10}C_2\,{}^{40}C_3$

$\therefore$ Total number of ways, if not more than 2 are defective $= {}^{40}C_5 + {}^{10}C_1\,{}^{40}C_4 + {}^{10}C_2\,{}^{40}C_3$

Its probability $= \left[{}^{40}C_5 + {}^{10}C_1\,{}^{40}C_4 + {}^{10}C_2\,{}^{40}C_3\right] / \left[{}^{50}C_5\right]$

$= \underline{\underline{0.9517}}$

COMBINATORIAL ANALYSIS

2-13

A family moves to a new city and requires the services of a doctor and a dentist. Two medical clinics are available. Each clinic has 2 doctors and 3 dentists. Both are to be selected from the same clinic. How many different choices are there for selecting one of each?

****

Using a tree diagram to enumerate the possible choices,

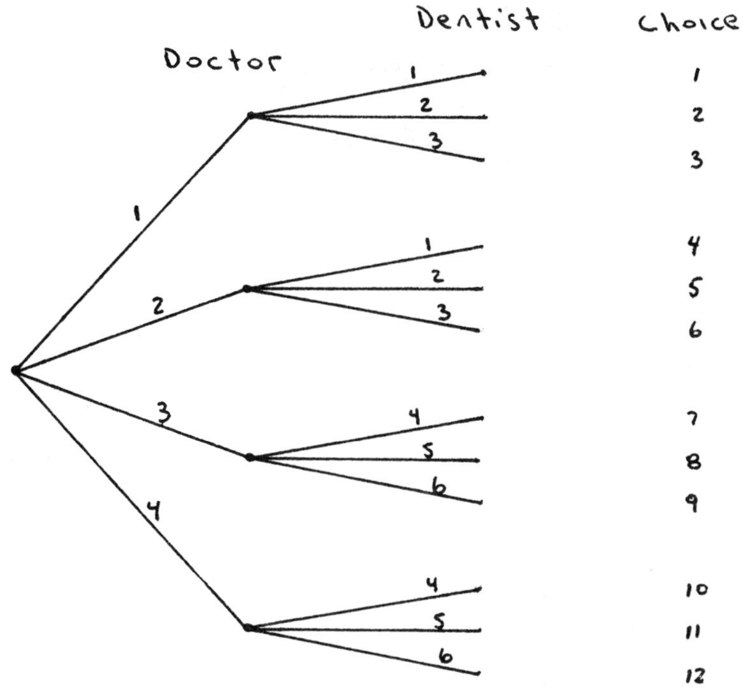

Ans. 12

2-14

A student engineers club is ready to elect four officers for the next academic year. At least three of the officers must be seniors. If there are 5 seniors out of 25 students in the club, and any club member may hold any office, in how many ways could a slate of officers be selected?

****

A selection (combination) of at least 3 of the 5 available seniors to serve on a 4-member slate of officers is given by

$$\binom{5}{3}\binom{20}{1} \quad + \quad \binom{5}{4}\binom{20}{0}$$

3 of the 5 seniors AND 1 of the 20 non-seniors OR 4 of the 5 seniors AND none of the 20 non-seniors

Once one of these selections is made, the students chosen may be permuted among any of the 4 positions, in 4! ways.

The desired number of possible slates of officers is thus

$$\left[\binom{5}{3}\binom{20}{1}+\binom{5}{4}\binom{20}{0}\right]\cdot 4! = \left[\frac{5!}{2!\,3!}\cdot\frac{20!}{19!\,1!}+\frac{5!}{1!\,4!}\cdot\frac{20!}{20!\,0!}\right]\cdot 4!$$

$$= \underline{\underline{4{,}920}}$$

2-15

How many different three letter fraternity and sorority names can be formed from the 24 letters of the Greek alphabet if no letter is used more than once?

How many if there is no restriction on duplication of letters?

****

a) Since arrangement of the letters matters $(\alpha\beta\gamma) \neq (\beta\alpha\gamma)$ the permutation of 24 letters taken 3 at a time is appropriate.

$$P_r^n = \frac{n!}{(n-r)!}$$

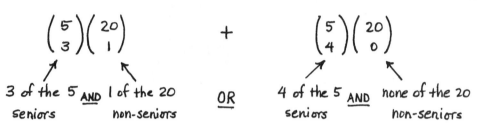

$$P_3^{24} = \frac{24!}{21!} = \frac{24 \cdot 23 \cdot 22 - (21!)}{21!} = \underline{12,144 \text{ names}}$$

b) There are 24 choices for the first letter and for each of these there are 24 choices for the second letter. For the third letter there are again 24 choices for each of the previous two letter arrangements

$$24 \cdot 24 \cdot 24 = \underline{13,824 \text{ different names}}$$

2-16

An urn contains 8 white balls mixed with 4 red balls. If two balls are drawn at random, determine the probability that one is a white ball and the other one a red ball.

****

If P stands for probability, then

P(1 white ball and one red ball drawn) =

$$\frac{\binom{8}{1}\binom{4}{1}}{\binom{12}{2}} = \frac{\dfrac{8!}{7!\,1!}\;\dfrac{4!}{3!\,1!}}{\dfrac{12!}{10!\,2!}} = \frac{8 \times 4}{\dfrac{12 \times 11}{2}}$$

$$= \frac{16}{33}$$

2-17

There are 8 graders for an exam which has 4 problems. The professor will select a different grader for each problem. How many ways can he choose graders for the exam? Suppose the professor only picks the 4 graders and leaves it up to them to decide who grades each problem. How many choices does he have now?

****

In the first part, the sample space of all possible outcomes includes all 4-tuples defined by

$$S = \left\{ (G_i^1, G_j^2, G_k^3, G_l^4) \;\middle|\; \begin{array}{l} i,j,k,l = 1,2,\ldots,8 \text{ and} \\ \text{all different} \end{array} \right\}$$

where G_i^1 is grader for Prob 1
G_j^2 is grader for Prob 2
etc

Total number of grader selections is given by

$$_8P_4 = \frac{8!}{4!} = 1680 \text{ permutations}$$

Alternatively, $\underset{\text{Prob 1}}{8 \text{ choices}} \times \underset{\text{Prob 2}}{7} \times \underset{\text{Prob 3}}{6} \times \underset{\text{Prob 4}}{5} = 1680$

In the second part, $(G_1^1, G_2^2, G_3^3, G_4^4)$, $(G_1^1, G_2^2, G_4^3, G_3^4)$, $(G_1^1, G_3^2, G_2^2, G_4^4)$, etc. are all permutations of the same combination, i.e. the first 4 graders.

Total number of selections is reduced to

$$_8C_4 = \frac{8!}{4!4!} = 70$$

Note, $_8C_4 = \frac{_8P_4}{4!}$ since there are 4! different permutations corresponding to the same combination.

2-18

A group of nine engineers in a department of a firm contain 4 who are single, 3 who are married, and 2 who are divorced. Three of the engineers are to be selected for advancement. Let n_1 denote the number of single engineers, n_2 the number of married engineers, and n_3 the number of divorced engineers among the three sected for advancement. Assuming that the three are randomly selected from the nine available, find the joint probability distribution for n_1 and n_2.

****

ALL POSSIBLE $P\{x_1, x_2\}$ VALUES ARE NEEDED, I.E.,

$$P\{x_1 = 0 \cap x_2 = 1\} = P\{0 \text{ SINGLE, } 1 \text{ MARRIED, } 2 \text{ DIVORCED}\}$$

$$= \frac{\binom{4}{0}\binom{3}{1}\binom{2}{2}}{\binom{9}{3}} = \frac{3}{84},$$

$$P\{x_1 = 0 \cap x_2 = 2\} = \frac{\binom{4}{0}\binom{3}{2}\binom{2}{1}}{\binom{9}{3}} = \frac{6}{84},$$

$$P\{x_1 = 0 \cap x_2 = 3\} = \frac{\binom{4}{0}\binom{3}{3}\binom{2}{0}}{\binom{9}{3}} = \frac{1}{84},$$

$$P\{x_1 = 1 \cap x_2 = 0\} = \frac{\binom{4}{1}\binom{3}{0}\binom{2}{2}}{\binom{9}{3}} = \frac{4}{84}, \text{ ETC.}$$

x_1 \ x_2	0	1	2	3
0	0	$\frac{3}{84}$	$\frac{6}{84}$	$\frac{1}{84}$
1	$\frac{4}{84}$	$\frac{24}{84}$	$\frac{12}{84}$	0
2	$\frac{12}{84}$	$\frac{18}{84}$	0	0
3	$\frac{4}{84}$	0	0	0

PROBABILITY OF EVENTS

2-19

A consulting firm specializing in the design of offshore structures plans to hire five new structural systems specialists from a pool of 18 recent M.S. graduates. Six of the 18 received their degree from the Massachusetts Institute of Technology (MIT), five from the California Institute of Technology, four from Georgia Institute of Technology and the other three from Virginia Polytechnic Institute (VPI). If the firm decides that all 18 applicants are equally qualified and therefore will select the potential recruits randomly from the pool, what is the probability that:

a) All five selected are MIT graduates.

b) No VPI graduates are selected

c) Two of the five selected are from VPI, one is from Georgia Tech and two from Cal-Tech.

****

a) proba (all 5 selected are from MIT) =

$$\frac{\binom{6}{5}}{\binom{18}{5}} = 0.0007$$

b) proba (no VPI grads. selected) = $$\frac{\binom{3}{0}\binom{15}{5}}{\binom{18}{5}} = 0.350$$

c) proba (2 VPI, 1 G-Tech, 2 Cal-Tech) =

$$\frac{\binom{3}{2}\binom{4}{1}\binom{5}{2}\binom{6}{0}}{\binom{18}{5}} = 0.014$$

COMPLEMENTARY EVENTS

2-20

The following information is known regarding three events A, B and C.

$p(A \cup B) = 0.75$ $\quad p(A \cap B) = 0.50$ $\quad p(B|C) = 0.50$

$P(B \cup C) = 0.75$ $\quad p(A \cap C) = 0.25$ $\quad p(A|B) = 0.75$

a) Find the following probabilities

i) $p(B)$
ii) $p(A)$
iii) $p(B|A)$
iv) $p(C)$
v) $p(B \cap C)$
vi) $p(A^c \cup B^c)$
vii) $p(A^c|B)$
viii) $p(B^c|A)$

b) Please answer the following questions and justify your answer:

i) Are events B and C mutually exclusive?
ii) Are events B and C independent?

****

a) i) $p(B) = \dfrac{p(A \cap B)}{p(A|B)} = \dfrac{0.5}{0.75} = \dfrac{2}{3} = 0.67$

ii) $p(A \cup B) = p(A) + p(B) - p(A \cap B)$,

therefore $p(A) = p(A \cup B) - p(B) + p(A \cap B) = 0.75 - 0.67 + 0.5$

$= 0.58$ (or $7/12$)

iii) $p(B|A) = \dfrac{p(A \cap B)}{p(A)} = \dfrac{0.5}{0.58} = 0.86$

iv) & v) $p(C)$ and $p(B \cap C)$ can be determined by solving the following 2 equations to 2 unkowns:

① $p(B \cap C) = p(C) \cdot p(B|C) = 0.5\,p(C)$

② $p(B \cup C) = p(B) + p(C) - p(B \cap C)$

$0.75 = 0.67 + p(C) - 0.5\,p(C)$

$\Rightarrow p(C) = 1/6 = 0.17$ and $p(B \cap C) = 1/12 = 0.08$

vi) $p(A^c \cup B^c) = p(A \cap B)^c$ [de Morgan's law]

$= 1 - p(A \cap B) = 1 - 0.5 = 0.50$

vii) $p(A^c|B) = 1 - p(A|B)$

$= 1 - 0.75 = 0.25$

viii) $p(B^c|A) = 1 - p(B|A)$

$= 1 - 0.86 = 0.14$

b) i) No ; $p(B \cap C) = 0.08 \neq 0 \Rightarrow B \cap C \neq \{\emptyset\}$

ii) No; $p(B|C) = 0.50$

$p(B) = 0.67$

$\} \; p(B|C) \neq p(B)$

CONDITIONAL PROBABILITY

2-21

A building contractor has an urgent electrical wiring job to complete on one of his sites. He usally subcontracts electrical work to one of two small firms. The probability that subcontractor A will be able to perform the job at such short notice is equal to 0.65, while for subcontractor B the probability is equal to 0.7. The probability that neither subcontractor is available for this job is equal to 0.2.

a) Calculate the probability that both subcontractors are available for this job.

b) Calculate the probability that at least one of the two is available for this job.

c) Calculate the probability that only one of the two is available for this job.

d) Calculate the conditional probability that subcontractor B is available given that the other subcontractor is unable to take on the job.

****

Let event $A = \{\text{subcontractor } A \text{ is available}\}$; $p(A) = 0.65$
event $B = \{\text{subcontractor } B \text{ is available}\}$; $p(B) = 0.70$

a) $p(A \cap B) = p(A) + p(B) - p(A \cup B)$

note: $p(A \cap B)$ = probability that both subcontractors are available.

$$p(A \cup B) = 1 - p(A \cup B)^c = 1 - p(A^c \cap B^c),$$

where $p(A^c \cap B^c)$ is the probability that neither subcontractor is available for this job, and is equal to 0.2.

Thus, $p(A \cup B) = 1 - 0.20 = 0.80$

and $p(A \cap B) = 0.65 + 0.70 - 0.80 = \underline{\underline{0.55}}$

b) proba (at least one is available) $= p(A \cup B) = \underline{\underline{0.80}}$ (from part(a))

c) proba (only one is available) $= p\left([A^c \cap B] \cup [A \cap B^c]\right)$

$$\begin{aligned}
&= p(A^c \cap B) + p(A \cap B^c) \\
&= p(A^c | B)p(B) + p(B^c | A)p(A) \\
&= p(B) - p(A \cap B) + p(A) - p(A \cap B) \\
&= p(A) + p(B) - 2p(A \cap B) = 0.70 + 0.65 - 2(0.55) \\
&= \underline{\underline{0.25}}
\end{aligned}$$

d) proba (B available given that A unavailable)

$$= p(B|A^c) = \frac{p(B \cap A^c)}{p(A^c)}$$

$$= \frac{p(B) - p(A \cap B)}{1 - p(A)}$$

$$= \frac{0.70 - 0.55}{0.35} = \frac{0.15}{0.35} = \frac{3}{7} = \underline{\underline{0.428}}$$

2-22

Salesman Joe visits the following cities to work

City	Population (in millions)
A	2
B	3
C	4

On his day off Joe resides in his home town, City D. On any given day there is a 10% probability that Joe will be off. If he is working, the probability of his being located in City A,B or C is proportional to the population. Find the probability that on a given day Joe is not in City B.

****

Define the following events: W = (Joe is working)

A = (Joe is in City A on a given day)

B = (City B) C = (City C) D = (City D)

$$Pr(B) = Pr(B|W)\,Pr(W) + Pr(B|\bar{W})\,Pr(\bar{W})$$

where $\bar{W}$ is the event Joe is not working

From the problem statement, the necessary conditional and unconditional probabilities are

$Pr(B|W) = 3/9$ $Pr(W) = 0.9$

$Pr(B|\bar{W}) = 0$ $Pr(\bar{W}) = 0.1$

Substituting in the above values yields

$$Pr(B) = (3/9)(0.9) + (0)(0.1) = 0.3$$

The probability of not being in City B is obtained from

$$Pr(\bar{B}) = 1 - Pr(B) = 1 - 0.3 = 0.7$$

2-23

A box of 500 rivets contains good rivets as well as rivets with the defects summarized below. If a rivet with a type A defect is obtained, what is the probability that this rivet also has a type B defect?

30 rivets with type A defects
15 rivets with type B defects
10 rivets with type C defects
4 rivets with type A and type B defects
3 rivets with type A and type C defects
2 rivets with type B and type C defects
1 rivet with all three types of defects

****

$$\text{Probability } B/A = \frac{P(AB)}{P(A)} = \frac{4/500}{30/500} = \frac{4}{30} = \underline{0.133}$$

2-24

On this exam 60% of the class will search through the text looking for answers. The remaining 40% are well prepared and do not need the text. Students not using the text get the correct answers 75% of the time, while those using the text get the correct answers only 25% of the time. Find the probability that

a) the first exam graded will have the correct answers.
b) a person who got the correct answers used the text.

****

Define events A_1 = (Use Text)
A_2 = (Don't Use Text)
B = (Get Correct Answers)

a)
$$\begin{aligned} Pr(B) &= Pr(B|A_1)Pr(A_1) + Pr(B|A_2)Pr(A_2) \\ &= (0.25)(0.60) + (0.75)(0.4) \\ &= 0.45 \end{aligned}$$

b)
$$Pr(A_1|B) = \frac{Pr(A_1 B)}{Pr(B)} = \frac{Pr(B|A_1)Pr(A_1)}{Pr(B)}$$

$$= \frac{(0.25)(0.6)}{0.45} = 0.33$$

2-25

Freshman at a university are given an entrance exam and only 65% of them pass. Of those who pass, 75% will graduate 4 years later. If 70% of the entering freshmen graduate, find the probability that a freshman who failed the exam will still graduate.

****

Define the events G = (Freshman will graduate)
P = (Freshman will pass exam)
F = (Freshman will fail exam)

$\Pr(G|F)$ is required. The unconditional probability of graduating is given by

$$\Pr(G) = \Pr(G|P)\Pr(P) + \Pr(G|F)\Pr(F)$$

From the statement of the problem, the following probabilities are known

$$\Pr(P) = 0.65 \implies \Pr(F) = 0.35$$
$$\Pr(G|P) = 0.75$$
$$\Pr(G) = 0.7$$

$$0.7 = (0.75)(0.65) + \Pr(G|F)(0.35)$$

$$\Pr(G|F) = \frac{0.7 - (0.75)(0.65)}{0.35} = 0.607$$

2-26

GIVEN A BOX CONTAINING 3 PENNIES, 1 NICKEL, 4 DIMES AND 2 QUARTERS. TWO COINS ARE DRAWN OUT, ONE AFTER THE OTHER.

IF THE FIRST COIN IS A DIME, WHAT IS THE PROBABILITY THAT THE SUM OF THE TWO COINSIS GREATER THAN $ 0.20 .

****

E_1 ----> FIRST COIN IS A DIME

FOR THE SUM TO BE GREATER THAN $ 0.20, THE SECOND COIN MUST BE A QUARTER.

P(Q/D) = 2/9 = 0.222

2-27

A bowl contains 7 black balls and 12 white balls. One ball is drawn at random without replacement. Then a second ball is drawn at random. What is the probability that both are black balls?

****

This is a case of conditional probability. If P stands for probability, then,

P(first ball is black and Second ball is black)

= P(first ball is black) X P(Second ball is black, given that the first one is black)

$$= \frac{7}{19} \times \frac{6}{18} = \frac{7}{57}$$

2-28

Each time a fair coin falls heads B pays A \$1, and each time it falls tails A pays B \$1. They start with \$2 each, and continue playing until one of them is broke. What is the probability that A wins, given that the first toss is heads?

****

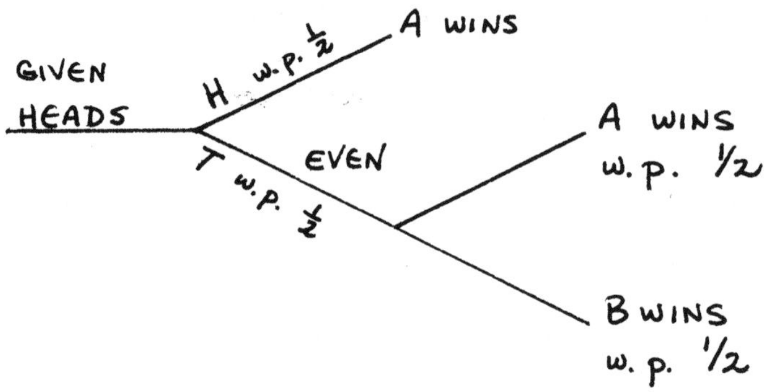

SINCE THE COIN IS FAIR, GIVEN THAT THE FIRST TOSS IS HEADS, WITH PROBABILITY 1/2 ANOTHER HEAD APPEARS SO THAT A WINS OR A TAIL APPEARS SO THAT A AND B ARE EVEN. AT ANYTIME WHEN A AND B ARE EVEN, EACH HAVE A PROBABILITY OF 1/2 OF WINNING, SINCE THE GAME IS FAIR.

$$
\begin{aligned}
Pr\{A \text{ WINS} \mid 1^{st} = H\} &= Pr\{A \text{ WINS AND } 2^{nd} = H \mid 1^{st} = H\} \\
&\quad + Pr\{A \text{ WINS AND } 2^{nd} = T \mid 1^{st} = H\} \\
&= Pr\{2^{nd} = H\} + Pr\{2^{nd} = T\} \times Pr\{A \text{ WINS}\} \\
&= 1/2 + (1/2 \times 1/2) = 3/4
\end{aligned}
$$

MULTIPLICATIVE RULE FOR INTERSECTIONS

2-29

The probability of the Campus Computer working when you want to do your homework is 0.95. The probability of all the terminals being busy when you need a terminal is 0.85. What is the probability of being able to do your homework on the computer when you are ready, willing and able?

****

Probability = (Prob. computer working)(Prob. terminal available)

$= (0.95)(1-0.85) = (0.95)(0.15) = \underline{0.1425}$

INDEPENDENCE

2-30

The probability that a man will be alive in 25 years is 3/5 and the probability his wife will be alive in 25 years is 2/3. Find the probability: a) both will be alive, b) only the man will be alive, c) only the wife will be alive, d) at least one will be alive, in 25 years. Assume independence for the events.

****

BY A TREE:

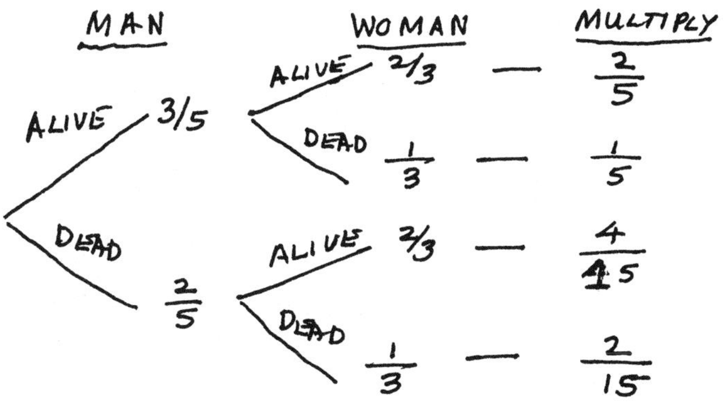

(a) ∴ P(BOTH ALIVE) = $2/5$

(b) P(ONLY MAN ALIVE) = $1/5$

(d) P(AT LEAST ONE ALIVE) $= 1 - P(\text{BOTH DEAD}) = 1 - \frac{2}{15}$

(c) P(ONLY WIFE) $= \frac{4}{15}$

$= \frac{13}{15}$

2-31

Of 250 professional employees in an international Transportation Systems Engineering firm, 130 hold a college degree at the Masters (or higher) level. Of the 180 Civil Engineering graduates working for the company, 100 hold a Masters' degree. The remaining 70 employees (without a Civil Engineering degree) have college degrees in areas such as urban planning, economics, computer science, operations research and management.

a) Calculate the probability that a randomly selected professional in this company holds a Masters degree and is not a Civil Engineering graduate.

b) If we meet a non-CE graduate at this company, calculate the conditional probability that he/she does not hold a Masters' degree.

c) Determine mathematically (from the data) whether being a "Civil Engineering graduate" and "holding a Masters' degree" are independent.

****

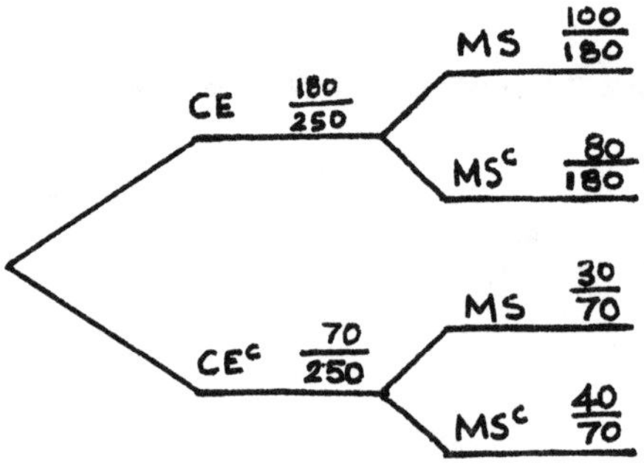

a) $p(MS \cap CE^c)$

$= p(MS|CE^c) \cdot p(CE^c)$

$= \frac{30}{70} \times \frac{70}{250} = \frac{3}{25}$

$= 0.12$

b) $p(MS^c \mid CE^c) = \frac{40}{70} = 0.571$

c) $p(CE) = \frac{180}{250}$; $p(MS) = \frac{130}{250}$

$p(CE) \cdot p(MS) = \frac{180}{250} \times \frac{130}{250} = 0.374$

$\neq$

$p(CE \cap MS) = p(MS|CE) \cdot p(CE)$

$= \frac{100}{180} \times \frac{180}{250} = \frac{100}{250} = 0.40$

since $p(CE \cap MS) \neq p(CE)\,p(MS)$,

then CE and MS are <u>not</u> independent.

2-32

a) The probability that a grader will make a marking error on any particular question of a multiple choice exam is 0.1. If there are 20 questions, and questions are graded independently, what is the probability that at least one error is made (on a given exam paper)?

b) Under the above assumptions, compute the probability that the first error made by the grader on a given exam occurs on the fifth question.

****

a) $p(\text{at least 1 error}) = 1 - p(\text{no errors})$

Letting A_i denote the event "grader makes an error on question i", $i=1,\ldots,20$; $p(A_i) = 0.1$

$$p(\text{no errors}) = p(A_1^c \cap A_2^c \cap \ldots \cap A_{20}^c)$$

$$= p(A_1^c)\,p(A_2^c) \cdots p(A_{20}^c) \quad [\text{by independence}]$$

$$= (0.9)^{20}$$

$$\therefore p(\text{at least one error}) = 1-(0.9)^{20} = \underline{\underline{0.878}}.$$

b) $p(1^{st}\text{ error occurs on fifth question})$

$$= p(\underbrace{A_1^c \cap A_2^c \cap A_3^c \cap A_4^c}_{\text{no errors on first four questions}} \cap \underbrace{A_5}_{\text{error on fifth question}})$$

$$= p(A_1^c)\,p(A_2^c)\,p(A_3^c)\,p(A_4^c)\,p(A_5)$$

$$= (0.9)^4 \times 0.1 = \underline{\underline{0.066}}.$$

2-33

A certain warning system consists of a detection camera and a computer such that the system fails if either one of these components fails. If the operational probability of detection camera is 0.85 for 90 days without failure and that of the computer 0.92 for 90 days without failure, determine the operating probability of the combined system for the 90 days period without failure. Consider failure of each component to be an independent event.

****

This is a case of independent events.

Probability of the combined system to function without failure for 90 days =

P(combined system) = P(detection camera) × P(computer)

$= 0.85 \times 0.92 = 0.782$

$\simeq 78.2\%$

2-34

A box contains 9 tickets numbered from 1 to 9 inclusive. If 3 tickets are drawn from the box one at a time without replacement, find the probability they are alternately either odd, even, odd or even, odd, even.

****

there are 5 odd and 4 even numbers in box:

$\therefore$ P (ODD–EVEN–ODD) OR (EVEN–ODD–EVEN)

$$= \frac{5}{9}\cdot\frac{4}{8}\cdot\frac{4}{7} + \frac{4}{9}\cdot\frac{5}{8}\cdot\frac{3}{7} = \frac{35}{126} = \frac{5}{18}$$

BAYES' RULE

2-35

A FIRM RENTS CARS FROM THREE RENTAL AGENCIES: 60% FROM AGENCY D, 20% FROM AGENCY E AND THE REST FROM F.

IF 12% OF THE CARS FROM D HAVE BAD TIRES AND 4% FROM E ARE BAD AND F HAS 10% BAD, WHAT IS THE PROBABILITY THAT A CAR THAT IS RENTED WILL HAVE BAD TIRES ?

****

THIS IS A PROBLEM INVOLVING BAYES' EQUATION

$B \rightarrow$ EVENT : BAD TIRES

$P(D) = 0.6 \quad P(E) = 0.2 \quad P(F) = 0.2$

(PROBABILITIES OF RENTING FROM AGENCIES)

$P(B/D) = 0.12 \quad P(B/E) = 0.04 \quad P(B/F) = 0.10$

$$P(B) = P(D)\cdot P(B/D) + P(E)\cdot P(B/E) + P(F)\cdot P(B/F)$$

$$= (.6)(.12) + (.2)(.04) + (.2)(.10) = 0.10$$

2-36

A group of students consists of 40% graduate students and 60% undergraduate students. If 20% of the graduate students and 40% of the undergraduates smokes, what is the probability a smoker is a graduate student?

****

BY A TREE;

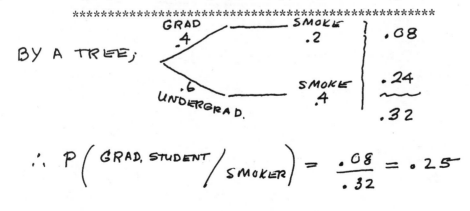

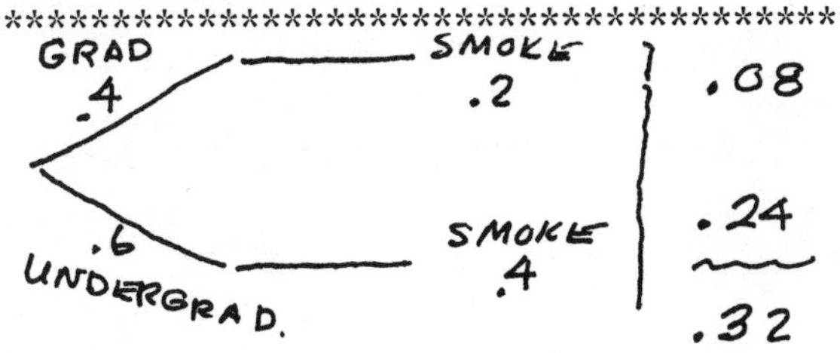

$$\therefore P\left(\text{GRAD. STUDENT}/\text{SMOKER}\right) = \frac{.08}{.32} = .25$$

2-37

An industrial psychologist has developed a test to identify people with a potential to be good managers. His statistics indicate that 30 percent of the employees in the company have the potential to be good managers. When his test is given to a large number of employees similar to the ones in his company 80 percent of the potentially good managers will pass the test while only 20 percent of the potentially poor managers will pass it.

a) What is the probability that an employee picked at random from the company will be a potentially good manager?

b) What is the probability that an employee picked at random from the company will pass the test.

c) What is the probability that an employee who passes the test will be a good manager?

****

P(Good Mgr.) = .30
P(Pass test) = .20
P(Pass/Good mgr.) = .80

P(poor mgr.) = .70
P(pass | poor mgr.) = 1 − .80 = .20

a) P(Good Mgr.) = .30

b) P(Pass test) = .20

c) P(Good Mgr./pass test) = Bays formula

$$P(E_1 | E_2) = \frac{P(E_1)\, P(E_2 | E_1)}{P(E_2)}$$

$$= \frac{(.30)(.80)}{(.30)(.80) + (.70)(.20)} = \frac{.24}{.24 + .14}$$

P(Good Mgr. | Pass test) = .63

2-38

A well-known problem in traffic engineering is that of the "dilemma zone" associated with drivers who see the traffic signal indication change from green to amber (before the red indication) as they are approaching the intersection. For drivers who are within 400 feet of the stop bar (at the intersection) at the onset of amber, it was found that, initially, 45% of drivers decelerate, 30% of drivers accelerate and the remaining 25% maintain constant speed. Of those who decelerate, 95% eventually come to a complete stop, while the remaining 5% continue through the intersection. For those who choose to accelerate, none come to a complete stop, while 30% of those who initially maintain constant speed ultimately come to a complete stop.

a) What is the probability that the next approaching driver at the onset of amber will decelerate and continue through the intersection without stopping?

b) What is the probability that the next approaching driver at the onset of amber will continue through the intersection without stopping?

c) Given that the last approaching driver came to a complete stop after the onset of amber, what is the conditional probability that he/she initially decelerated?

****

We define the following events:
$A = \{\text{driver accelerates}\}$
$B = \{\text{driver decelerates}\}$
$C = \{\text{driver maintains constant speed}\}$
$S = \{\text{driver stops at intersection}\}$

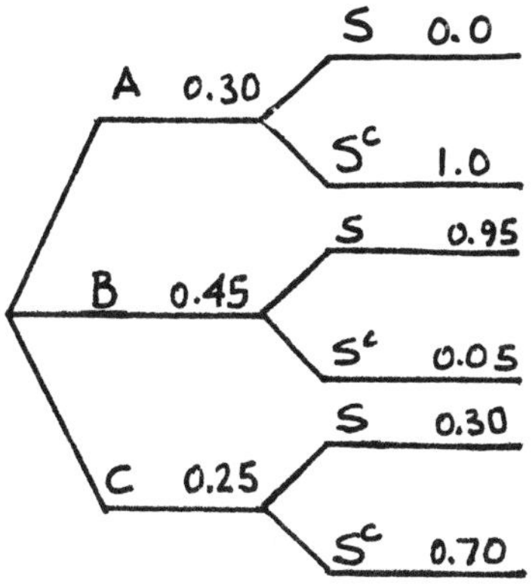

a) $p(B \cap S^c) = p(S^c|B) \times p(B)$

$= 0.05 \times 0.45 = 0.0225$, or $\underline{\underline{2.25\%}}$

b) $p(S^c) = p\big([A \cap S^c] \cup [B \cap S^c] \cup [C \cap S^c]\big)$

$= p(A \cap S^c) + p(B \cap S^c) + p(C \cap S^c)$

$= 0.3 \times 1.0 + 0.45 \times 0.05 + 0.25 \times 0.70 = 0.4975$, or $\underline{\underline{49.75\%}}$

c) $p(B|S) = \dfrac{p(B \cap S)}{p(S)}$

$= \dfrac{p(S|B)\,p(B)}{p(S|A)\,p(A) + p(S|B)\,p(B) + p(S|C)\,p(C)}$ (BAYES' RULE)

$= 0.8507$, or $\underline{\underline{85.07\%}}$

2-39

Ford Motor Company manufactures trucks in three plants, say, A, B, and C. On the average, 4 trucks out of 500 assembled at A must be recalled, 10 out of 800 assembled at B must be recalled, and 10 out of 1000 assembled at C must be recalled. If a customer purchases a truck which is recalled from a dealer which receives 30%, 40%, and 30% of its trucks from plants A, B, and C, respectively, what is the probability that the truck came from plant (a) A, (b) B, or (c) C?

****

LET A, B, C STAND FOR THE EVENT THAT A TRUCK COMES FROM PLANT A, B, OR C, RESPECTIVELY, AND LET R BE THE EVENT OF A RECALL.

$$\Pr\{A|R\} = \frac{\Pr\{A \text{ and } R\}}{\Pr\{R\}} = \frac{\Pr\{R|A\}\Pr\{A\}}{\Pr\{R\}}$$

$$= \frac{\Pr\{R|A\}\Pr\{A\}}{\Pr\{R|A\}\Pr\{A\} + \Pr\{R|B\}\Pr\{B\} + \Pr\{R|C\}\Pr\{C\}}$$

$$= \frac{(4/500)(3/10)}{(4/500)(3/10) + (10/800)(4/10) + (10/1000)(3/10)} \quad \text{(BAYES' THEOREM)}$$

$$\doteq 0.2308$$

$$\Pr\{B|R\} = \frac{\Pr\{R|B\}\Pr\{B\}}{\Pr\{R\}}$$

$$= \frac{(10/800)(4/10)}{0.0104}$$

$$= 0.4808$$

NOTE: AS A CHECK, OBSERVE THAT THE 3 PROBABILITIES SUM TO 1.0.

$$\Pr\{C/R\} = \frac{\Pr\{R|C\}\Pr\{C\}}{\Pr\{R\}}$$

$$= \frac{(10/1000)(3/10)}{0.0104}$$

$$= 0.2884$$

2-40

If at Tundra U. P(A|M)=0.02 of the men, and P(A|W)=0.005 of the women are taller than 6 feet, with the fraction of women P(W)=0.60, a student is selected at random who is taller than 6 feet tall, what is the probability P(W|A) that the student is a woman?

****

$$P(W/A) = \frac{P(A|W)P(W)}{P(A|W)P(W)+P(A|M)P(M)}$$

$$P(A|W) = 0.005, \quad P(W) = 0.6$$

$$P(A|M) = 0.02, \quad P(M) = 0.4$$

$$\therefore \quad P(W/A) = \frac{0.005(0.6)}{0.005(0.6) + 0.02(0.4)}$$

$$P(W/A) = 0.2727$$

FUNDAMENTAL PROBABILITY LAWS

2-41

A dresser drawer contains 17 identical black socks and 13 identical red socks. A student who dresses in the dark chooses 2 socks at random from the drawer. What is the probability that he will have a matched pair of socks?

****

Matched pair = 2 black or 2 Red

$$P(B_1 \& B_2) = P(B_1)\,P(B_2 | B_1) = \left(\tfrac{17}{30}\right)\left(\tfrac{16}{29}\right) = .3126$$

$$P(R_1 \& R_2) = P(R_1)\,P(R_2 | R_1) = \left(\tfrac{13}{30}\right)\left(\tfrac{12}{29}\right) = .1793$$

P(matched pair = .49

2-42

Let A and B represent two events in a sample space with P(A) = 0.3 and P(A U B) = 0.8 . Find P(B) if A and B are independent events but not mutually exclusive. (Note that P(A) denotes the probability of event A occurring, and A U B denotes the union of the events A, B - A <u>or</u> B occurring.)

****

$$P(A \cup B) = P(A) + P(B) - P(A \cap B)$$

since A and B are not mutually exclusive ($A \cap B$ denotes the intersection of the events; in other words, A <u>and</u> B occur).

$$P(A \cup B) = P(A) + P(B) - P(A) \cdot P(B)$$

since A and B are independent.

Substituting known values,

$$0.8 = 0.3 + P(B) - (0.3) \cdot P(B)$$

$$0.8 = 0.3 + P(B) \cdot (1 - 0.3)$$

$$P(B) = \frac{0.8 - 0.3}{1 - 0.3}$$

$$P(B) = \frac{5}{7} \cong 0.7143$$

2-43

Let A and B be events with P(A)=1/2, P(B)=1/3, and P(AB)=1/4.

Determine the following:

(a) $P(A|B)$ (e) $P(\bar{A}B)$

(b) $P(B|A)$ (f) $P(\bar{A}|\bar{B})$

(c) $P(A \cup B)$ (g) $P(\bar{B}|\bar{A})$

(d) $P(A\bar{B})$ (h) $P(\overline{AB})$

****

(a) $P(A|B) = P(AB)/P(B) = \frac{1}{4}/\frac{1}{3} = \frac{3}{4}$

(b) $P(B|A) = P(AB)/P(A) = \frac{1}{4}/\frac{1}{2} = \frac{1}{2}$

(c) $P(A \cup B) = P(A) + P(B) - P(AB) = \frac{1}{2} + \frac{1}{3} - \frac{1}{4} = \frac{2}{3}$

(d) $P(A) = P(AB) + P(A\bar{B}) \quad \therefore \quad P(A\bar{B}) = P(A) - P(AB)$

$P(A\bar{B}) = \frac{1}{2} - \frac{1}{4} = \frac{1}{4}$

(e) $P(B) = P(AB) + P(\bar{A}B) \quad \therefore \quad P(\bar{A}B) = P(B) - P(AB)$

$P(\bar{A}B) = \frac{1}{3} - \frac{1}{4} = \frac{1}{12}$

(f) $P(\bar{A}|\bar{B}) = P(\bar{A}\bar{B})/P(\bar{B}), \quad P(\bar{B}) = P(A\bar{B}) + P(\bar{A}\bar{B})$

$\therefore \quad P(\bar{A}\bar{B}) = P(\bar{B}) - P(A\bar{B}) = \{1 - P(B)\} - P(A\bar{B})$

$= \{1 - \frac{1}{3}\} - \frac{1}{4} = \frac{5}{12}$

$\therefore \quad P(\bar{A}|\bar{B}) = \frac{5}{12}/\frac{2}{3} = \frac{5}{8}$

(g) $P(\bar{B}|\bar{A}) = P(\bar{A}\bar{B})/P(\bar{A}) = \frac{5}{12}/\frac{1}{2} = \frac{5}{6}$

(h) $P(\overline{AB}) = P(\bar{A} \cup \bar{B}) = P(\bar{A}) + P(\bar{B}) - P(\bar{A}\bar{B})$

$= \frac{1}{2} + \frac{2}{3} - \frac{5}{12} = \frac{3}{4}$

OR $P(AB) + P(\overline{AB}) = 1 \quad \therefore \quad P(\overline{AB}) = 1 - P(AB)$

$P(\overline{AB}) = 1 - \frac{1}{4} = \frac{3}{4}$

2-44

Three students attempt independently to solve a difficult statistics problem. Their probabilities of success are 50%, 40% and 20%. What is the probability that:

a) none of them will solve the problem?

b) only one of them will solve the problem?

c) at least one of them will solve the problem?

****

a) $P(\text{none}) = (1-.5)(1-.4)(1-.2) = \underline{\underline{.24}}$

b) $P(\text{only one}) = .5(1-.4)(1-.2)+(1-.5)(.4)(1-.2)+(1-.5)(1-.4)(.2)$
$= .24 + .16 + .06 = \underline{\underline{.46}}$

c) $P(\text{at least one}) = 1-P(\text{none}) = 1-.24 = \underline{\underline{.76}}$

2-45

An assembly consists of three parts with the probability of a defective in the parts being 0.1, 0.2 and 0.25 resp. If these parts are assembled at random, what is the probability exactly one of the three parts in the assembly will be defective?

****

P(exactly one defective) = P(A is good; B, C bad) OR (B is good; A, C bad) OR (C is good; A, B bad)

$= (.1)(.8)(.75) + (.9)(.2)(.75) + (.9)(.8)(.25)$

$= .06 + .135 + .18 = .375$

2-46

Events A, B and C are associated with a particular experiment and sample space. Events A and B are mutually exclusive, while events A and C are independent. The probability that event B occurs given that event C occurs is 0.30 and the probability that event A occurs given that event C occurs is 0.25. Events A and B are equally likely and event C is three times as likely to occur as event A. What is the probability that none of the three events A, B and C occur?

****

$$P(A' \cap B' \cap C') = P(A \cup B \cup C)' = 1 - P(A \cup B \cup C)$$

$$P(A \cup B \cup C) = P(A) + P(B) + P(C) - P(A \cap B) - P(A \cap C) - P(B \cap C) + P(A \cap B \cap C)$$

$$P(A \cap B) = 0, \quad \therefore \quad P(A \cap B \cap C) = 0.$$

$$P(A|C) = P(A) = 0.25, \quad \therefore \quad P(B) = 0.25, \quad P(C) = 0.75$$

$$P(B \cap C) = P(B|C)P(C) = (0.30)(0.75) = 0.2250$$

$$P(A \cap C) = P(A|C)P(C) = (.25)(.75) = 0.1875$$

$$P(A \cup B \cup C) = .25 + .25 + .75 - 0 - .1875 - .2250 + 0 = .8375$$

$$P(A' \cap B' \cap C') = 1 - 0.8375 = 0.1625$$

2-47

An urn contains 3 balls bearing the numbers 1, 2, 3 respectively. Five balls are drawn with replacement. What is the probability that exactly 1 number occurs exactly 3 times?

****

FOR THE ① to occur 3 TIMES: $\binom{5}{3}\left(\frac{1}{3}\right)^3\left(\frac{2}{3}\right)^2 = \frac{40}{243}$

∴ FOR ANY ONE NUMBER TO OCCUR THREE TIMES:

$$\frac{40}{243} + \frac{40}{243} + \frac{40}{243} = \frac{40}{81} = .494$$

2-48

The probability of clear weather on a given day in N.Y. is 0.4 and L.A. is 0.3. The probability of clear weather in at least one of the two cities is 0.58.

a) Find the probability of clear weather in N.Y. and L.A. on the same day.
b) Are the two events independent?
c) Find the probability of neither city having clear weather the same day.
d) Find the probability of clear weather in either N.Y. of L.A., but not both, on the same day.

****

Let event A = (Clear weather in N.Y.) $P(A) = 0.4$
B = (Clear weather in L.A.) $P(B) = 0.3$

$$P(A+B) = 0.58$$

a)
$$P(AB) = P(A) + P(B) - P(A+B) = 0.4 + 0.3 - 0.58 = 0.12$$

b) A & B are independent if $P(AB) = P(A)P(B)$

In this example, the condition is satisfied and therefore A & B are statistically independent.

c) We are looking for $P(\bar{A}\bar{B})$. From a Venn diagram it is easy to verify that

$$P(\bar{A}\bar{B}) + P(A+B) = 1 \implies P(\bar{A}\bar{B}) = 1 - P(A+B) = 1 - 0.58 = 0.42$$

d) We are looking for $P(A\bar{B} + \bar{A}B)$

$$P(A\bar{B} + \bar{A}B) = P(A\bar{B}) + P(\bar{A}B) - P(A\bar{B}\,\bar{A}B)$$
$$= P(A)P(\bar{B}) + P(\bar{A})P(B)$$
$$= (0.4)(0.7) + (0.6)(0.3)$$
$$= 0.46$$

since A & B are independent and $A\bar{B}\bar{A}B$ is null event

Alternatively, $P(\text{Exactly } 1) + P(\text{Both}) + P(\text{Neither}) = 1$
$$P(\text{Exactly } 1) = 1 - P(\text{Both}) - P(\text{Neither}) = 1 - 0.12 - 0.42 = 0.46$$

2-49

In a certain developed country, 52% of the population is female. Records show that before the age of fifty, 15% of the male population will develop heart disease and 20% will develop arthritis. For females under fifty, 4% will develop heart disease and 35% arthritis. What is the probability that a person under fifty selected from the population known to have heart disease <u>or</u> arthritis is male? (Assume the two diseases are independent but not mutually exclusive.)

****

Let M represent male, F female, A arthritis, and H heart disease. Let C represent the event of having arthritis <u>or</u> heart disease (or both). Then $C = A \cup H$ (A union H), and

$$P(C) = P(A) + P(H) - P(A \cap H)^{*} \quad \text{since A and H are not mutually exclusive}$$
$$= P(A) + P(H) - P(A) \cdot P(H) \quad \text{since A and H are independent}$$

The conditional probability of having arthritis or heart disease (or both) <u>given that</u> the selected individual is male is given by

$$P(C/M) = 0.20 + 0.15 - (0.20)(0.15) = 0.320 \text{ . Similarly,}$$
$$P(C/F) = 0.35 + 0.04 - (0.35)(0.04) = 0.376 \text{ .}$$

It is also known that $P(M) = 0.48$ and $P(F) = 0.52$.

With this information, the desired conditional probability of selecting a male from the population given that the selected individual has arthritis or heart disease (or both) can be found as follows using Bayes' Theorem:

$$P(M/C) = \frac{P(C/M) \cdot P(M)}{P(C/M) \cdot P(M) + P(C/F) \cdot P(F)}$$

$$= \frac{(0.320)(0.48)}{(0.320)(0.48) + (0.376)(0.52)}$$

$$\cong \underline{\underline{0.44}}$$

*$A \cap H$ is A intersect H or A <u>and</u> H occur.

2-50

Consider a target practice by two varsity students, A and B. Let the Probability that A hits the target P(A) = 1/3, and the probability that B hits it P(B) = 1/4, and the probability of A or B hitting it P(AUB) = 1/2.

Find:
(a) $P(A|B)$
(b) $P(B|A)$
(c) $P(A|\bar{B})$.

****

(a) $P(A|B) = \dfrac{P(A \cap B)}{P(B)}$, $P(A \cup B) = P(A) + P(B) - P(A \cap B)$

$$P(A \cap B) = P(A) + P(B) - P(A \cup B) = \frac{1}{3} + \frac{1}{4} - \frac{1}{2} = \frac{1}{12}$$

$$\therefore P(A|B) = \frac{1/12}{1/4} = 1/3$$

(b) $$P(B|A) = \frac{P(A \cap B)}{P(A)} = \frac{1/12}{1/3} = 1/4$$

(c) $$P(A) = P(A \cap B) + P(A \cap \bar{B})$$

$$P(A \cap \bar{B}) = P(A) - P(A \cap B) = 1/3 - 1/12 = \frac{3}{12} = 1/4$$

$$\therefore P(A|\bar{B}) = \frac{P(A \cap \bar{B})}{P(\bar{B})} = \frac{1/4}{1 - 1/4} = 1/3$$

2-51

One bag contains four white balls and three black balls, and a second bag contains three white balls and five black balls. One ball is drawn from the first bag and placed unseen in the second bag. What is the probability that a ball now drawn from the second bag is black?

****

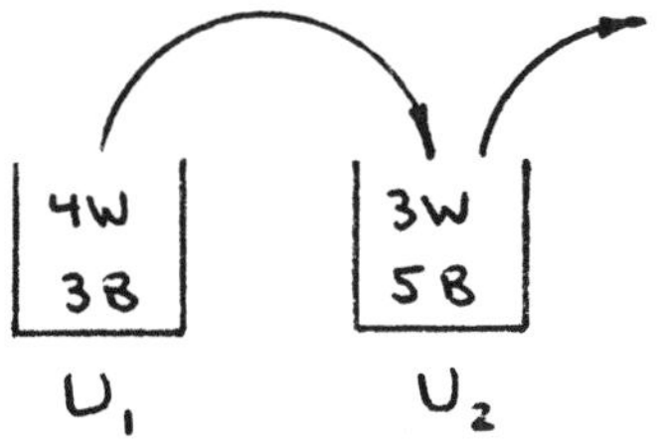

Define the following events

A = (Ball from U_1 to U_2 is white)
B = (Ball from U_1 to U_2 is black)
C = (Ball from U_2 is black)

The probability of event C occurring is conditional on whether event A or B occurred,

i.e.
$$\begin{aligned} Pr(C) &= Pr(C|A)\,Pr(A) + Pr(C|B)\,Pr(B) \\ &= (5/9)(4/7) + (6/9)(3/7) \\ &= 38/63 \end{aligned}$$

The solution can be easily visualized from a tree diagram.

$$Pr(C) = \frac{4}{7}\cdot\frac{5}{9} + \frac{3}{7}\cdot\frac{6}{9}$$

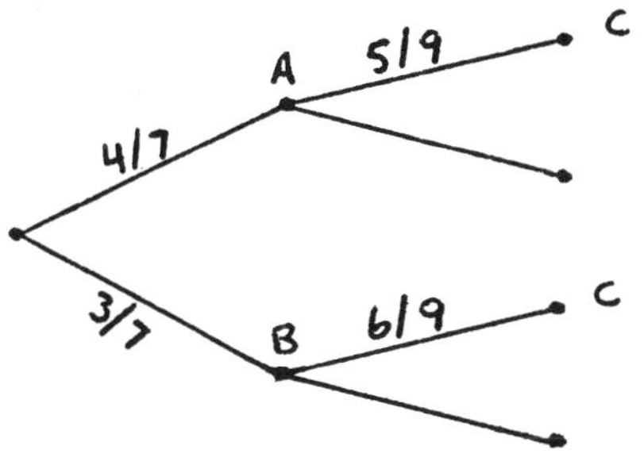

2-52

Each of two packages of six flashlight batteries contain exactly two in-operable batteries. a) If two batteries are selected from each package, what is the probability that all four batteries will function. b) Suppose two batteries were randomly selected from Pack #1 and mixed with those from Pack #2. Then two were randomly drawn from the eight in Pack #2. What is the probability that both will function?

****

(a) $P(2\text{ GOOD}) = \frac{4}{6}\cdot\frac{3}{5} = \frac{2}{5}$

$\therefore\ P(\text{ALL FOUR GOOD}) = \frac{2}{5}\cdot\frac{2}{5} = \frac{4}{25}$

(b) ON TRANSFERRING TWO BATTERIES,

Branch	Probability	RESULT
2 GOOD	$\frac{4}{6}\cdot\frac{3}{5} = \frac{2}{5}$	2B, 6G
1 GOOD, 1 BAD	$\frac{4}{6}\cdot\frac{2}{5} + \frac{2}{6}\cdot\frac{4}{5} = 8/15$	3B, 5G
2 BAD	$\frac{2}{6}\cdot\frac{1}{5} = 1/15$	4B, 4G

$\therefore\ P(2\text{ GOOD}) = \frac{2}{5}\left(\frac{6}{8}\cdot\frac{5}{7}\right) + \frac{8}{15}\left(\frac{5}{8}\cdot\frac{4}{7}\right) + \frac{1}{15}\left(\frac{4}{8}\cdot\frac{3}{7}\right)$

$= \frac{44}{105}$

2-53

A rental car facility has 10 foreign cars and 15 domestic cars waiting to be serviced on a particular Saturday morning. Because there are so few mechanics working on Saturday, only 6 cars can be serviced.

a) If the 6 cars are chosen at random, what is the probability that 3 of the cars selected are foreign and the other 3 domestic?

b) What is the probability that all of the selected cars are foreign?

c) What is the probability that at least 3 of the selected cars are domestic?

****

a) proba (3 Foreign and 3 Domestic) $= \dfrac{\binom{10}{3}\binom{15}{3}}{\binom{25}{6}}$

$$= \frac{\frac{10!}{7!\,3!}\;\frac{15!}{3!\,12!}}{\frac{25!}{6!\,19!}} = \underline{\underline{0.3083}}$$

b) proba (all 6 are Foreign) $= \dfrac{\binom{10}{6}\binom{15}{0}}{\binom{25}{6}} = \dfrac{\frac{10!}{6!\,4!} \times 1}{\frac{25!}{6!\,9!}} = \underline{\underline{0.0012}}$

c) proba (at least 3 Domestic)

$= 1 -$ proba (0 or 1 or 2 Domestic)

$$= 1 - \frac{\binom{10}{6}\binom{15}{0} + \binom{10}{5}\binom{15}{1} + \binom{10}{4}\binom{15}{2}}{\binom{25}{6}}$$

$= \underline{\underline{0.8530}}$

2-54

Given the following odds:

In favor of event A,	2:1
In favor of event B,	1:5
In favor of event A or B or both,	5:1

Find the probability of the event AB occurring.

****

$$Pr(A+B) = Pr(A) + Pr(B) - Pr(AB)$$

Using the given odds to find the probabilities,

$$Pr(A) = \frac{2}{3} \qquad Pr(B) = \frac{1}{6} \qquad Pr(A+B) = \frac{5}{6}$$

$$\frac{5}{6} = \frac{2}{3} + \frac{1}{6} - Pr(AB)$$

$$Pr(AB) = 0$$

3 DISCRETE RANDOM VARIABLES

PROBABILITY MASS FUNCTION

3-1

Suppose that a die is loaded in such a manner that for x = 1,2,3,4,5,6 the probability of the face marked x turning up when the die is tossed is proportional to x. Find the probability of an even number resulting when the die is tossed.

****

$f_X(x) \sim x$ for $x = 1,2,3,4,5,6$

The proportionality constant K in $f_X(x) = Kx$ is evaluated from

$$1 = \sum_{x=1}^{6} f_X(x) = \sum_{x=1}^{6} Kx = K(1+2+3+4+5+6)$$

$$1 = 21K$$

$$K = 1/21$$

$$\begin{aligned} Pr(\text{Even number}) &= Pr(X = 2 \text{ or } X=4 \text{ or } X=6) \\ &= Pr(X=2) + Pr(X=4) + Pr(X=6) \\ &= f_X(2) + f_X(4) + f_X(6) \\ &= \frac{2}{21} + \frac{4}{21} + \frac{6}{21} \\ &= \frac{12}{21} \end{aligned}$$

Note, the "events" (X=2), (X=4) and (X=6) are disjoint

3-2

Let X be the number of taxicabs that a customer finds waiting outside the Hyatt Regency hotel anytime between 5:00 PM and midnight on weeknights. The probability mass function for X is

$$f_X(x) = \frac{x}{15}, \quad x = 1,2,\ldots,5$$

a) Show that $f_X(x)$ is a valid probability mass function.

b) What is the probability that a randomly selected customer (between 5:00 - midnight) will find exactly 3 taxicabs waiting?

c) What is the probability that a randomly selected passenger will find at least 2 taxicabs but no more than 4 taxicabs waiting outside the hotel between 5:00 - 12:00 PM.?

****

a) ① $0 \leq f_X(x) \leq 1$ for $x = 1,2,\ldots,5$

② $\sum f_X(x) = \frac{1}{15} + \frac{2}{15} + \frac{}{15} + \frac{4}{15} + \frac{5}{15} = \frac{15}{15} = 1$

① & ② ⇒ $f_X(x)$ is a valid pmf.

b) $P(X=3) = f_X(3) = \frac{3}{15}$

c) $P(2 \leq X \leq 4) = f_X(2) + f_X(3) + f_X(4)$

$= \frac{2}{15} + \frac{3}{15} + \frac{4}{15} = \frac{9}{15} = \frac{3}{5}$

3-3

Suppose X is a random variable having the density f given by

x	-3	-1	0	1	2	3	5	8
f(x)	.1	.2	.15	.2	.1	.15	.05	.05

Find the following: a) P(X is negative), b) $P(X = -3/X \leq 0)$, c) $P(X \geq 3/X > 0)$, d) E(X), e) E(3x-5)

****

(a) P(X is negative) = .1 + .2 = .3

(b) $P(X=-3/X\le 0) = \frac{.1}{.1+.2+.15} = .1/.45 = 2/9$

(c) $P(X\ge 3/X>0) = \frac{.15+.05+.05}{.2+.1+.15+.05+.05} = \frac{.25}{.55} = \frac{5}{11}$

(d) $E(X) = \sum x\cdot f(x) = 1.00$

(e) $E(3X-5) = 3E(X)-5 = 3-5 = -2$

3-4

A random variable has the probability function p(x) = kx; x = 1,2,3,4,5. Find a) k, b) $p(x \geq 3)$, c) μ, d) σ^2.

****

(a) the probability distribution has the following form:

X_i	1	2	3	4	5
$p(x_i)$	k	$2k$	$3k$	$4k$	$5k$

Since $\sum p(x) = 1$, $k = 1/15$

(b) $p(x\ge 3) = 12k = 12/15 = 4/5$

(c) $\mu = \sum x_i \cdot p(x_i) = 1\cdot\frac{1}{15} + 2\cdot\frac{2}{15} + 3\cdot\frac{3}{15} + 4\cdot\frac{4}{15} + 5\cdot\frac{5}{15} = \frac{55}{15} = \frac{11}{3} = 3.67$

(d) $\sigma^2 = E(x^2) - \mu^2 = \sum x_i^2\cdot p(x_i) - \mu^2$

$E(x^2) = 1^2\cdot\frac{1}{15} + 2^2\cdot\frac{2}{15} + 3^2\cdot\frac{3}{15} + 4^2\cdot\frac{4}{15} + 5^2\cdot\frac{5}{15}$

$= 225/15 = 15$

then $\sigma^2 = 15 - \left(\frac{11}{3}\right)^2 = 14/9 = 1.56$

3-5

Consider the following probability distribution for the discrete random variable X.

x	1000	2000	3000	4000
f(x)	.018	.086	.474	.422

Determine the expected value and variance of X.

****

$$\mu_X = E(X) = (1000)(.018) + (2000)(.086) + (3000)(.474) + (4000)(.422)$$
$$= 3300$$

$$\sigma_X^2 = V(X) = E(X^2) - [E(X)]^2$$
$$E(X^2) = (1000)^2(.018) + \ldots + (4000)^2(.422)$$
$$= 11,380,000$$

$$\sigma_X^2 = 11,380,000 - (3300)^2 = 490,000$$

3-6

A process randomly generates digits 0, 1, 2, ... 9 with equal probabilities. If T is a random variable representing the sum of 20 digits taken from the process, find the mean & variance of T.

****

FOR PARENT DISTRIBUTION:

X	0	1	2	3	4	5	6	7	8	9
p(x)	ALL 0.1									

$$\mu = 4.5, \ \sigma^2 = 33/4$$

$$T = X_1 + X_2 + \ldots\ldots + X_{20}$$

AND $\mu_T = 20\,\mu_X = 20\,(4.5) = 90$

$$\sigma_T^2 = 20\,\sigma_X^2 = 20 \cdot \frac{33}{4} = 165$$

3-7

A con man tosses three coins simultaneously. One of the coins is two-headed, unknown to a passing sucker. The other two coins are fair. The con man will pay the sucker, he says, \$10 if three tails are tossed, and \$3 if three heads are tossed. The sucker must pay the con man \$2 if one head is tossed and \$3 if two heads are tossed. This seems like a good deal to the sucker, so he stays with the game through many tosses of the coins. What is the con man's expected profit on the game?

****

Since most of the payoffs are based on heads tossed, let the random variable x count the number of heads tossed on a given trial of the game. With fair coins, x could be 0, 1, 2, or 3 on a given toss. However, one of these coins is two-headed, so $P(x=0) = 0$. Letting H represent a head and T a tail, and remembering that one coin will <u>always</u> be a head, the remaining probabilities are:

$$P(x=1) = HTT = (1)(\tfrac{1}{2})(\tfrac{1}{2}) = \tfrac{1}{4}$$
$$P(x=2) = HHT \text{ or } HTH = (1)(\tfrac{1}{2})(\tfrac{1}{2}) + (1)(\tfrac{1}{2})(\tfrac{1}{2}) = \tfrac{1}{2}$$
$$P(x=3) = HHH = (1)(\tfrac{1}{2})(\tfrac{1}{2}) = \tfrac{1}{4}$$

All probabilities must sum to one, which they do $(0+\frac{1}{4}+\frac{1}{2}+\frac{1}{4}=1)$.

The con man's expected profit over many trials of the game is found by summing the products of the payoff (from the viewpoint of the con man) and associated probability of occurrence of that payoff, or

$$\text{Expected profit} = \underbrace{(-\$10)(0)}_{\text{no heads}} + \underbrace{(\$2)(\tfrac{1}{4})}_{\text{1 head}} + \underbrace{(\$3)(\tfrac{1}{2})}_{\text{2 heads}} + \underbrace{(-\$3)(\tfrac{1}{4})}_{\text{3 heads}}$$

$$= \underline{\underline{\$1.25}}$$

Advice: Don't enter the game!

3-8

A non profit lottery costs \$38.67 to play. A person draws a ticket from a box and wins an amount as shown in the table below.

Color	Number of tickets	Winnings
white	200	\$0
blue	75	\$10
black	25	\$100
red	15	?
brown	5	\$1000

Suppose you enter the lottery 3 times and get a red, white and blue ticket. What is your net loss or gain if you include the cost of playing?

****

The winnings for drawing a red ball are determined by setting the expected winnings equal to the price of a lottery ticket because the lottery is non-profit.

Let r.v. W = Winnings

$$Pr(W=w) = f_W(w) = \begin{cases} 200/320 & w = 0 \\ 75/320 & 10 \\ 25/320 & 100 \\ 15/320 & X \\ 5/320 & 1000 \end{cases}$$

$$\text{Expected Winnings} = E(W) = \sum_w w\, f_W(w)$$

$$E(W) = 0(200/320) + 10(75/320) + \text{--------} + 1000(5/320)$$

$$38.67 = \frac{750 + 2500 + 15X + 5000}{320}$$

$X = 274.96$ i.e. a red ball wins about \$275.

Net (Loss) or Gain from red, white and blue ticket

$= \$275 + \$0 + \$10 - 3(\$38.67)$

$= \$169$

3-9

Items coming off a production line are sampled randomly in samples of three items, with each item being classified as defective or nondefective. From past data it has been determined that one out of ten items is defective on average. If x represents a random variable counting the number of defectives in a given sample, develop the probability mass function of x and graph it.

****

A sample of three items may have 0, 1, 2, or 3 defectives, so x can assume any one of these values. The associated probabilities are determined from the binomial distribution (since an item is either defective or nondefective) as follows:

$$P(x=0) = \binom{3}{0}(0.1)^0(0.9)^3 = \frac{3!}{3!\,0!}(0.9)^3 = (0.9)^3 = 0.729$$

$$P(x=1) = \binom{3}{1}(0.1)^1(0.9)^2 = \frac{3!}{2!\,1!}(0.1)(0.9)^2 = 3(0.1)(0.9)^2 = 0.243$$

$$P(x=2) = \binom{3}{2}(0.1)^2(0.9)^1 = \frac{3!}{1!\,2!}(0.1)^2(0.9) = 3(0.1)^2(0.9) = 0.027$$

$$P(x=3) = \binom{3}{3}(0.1)^3(0.9)^0 = \frac{3!}{0!\,3!}(0.1)^3 = (0.1)^3 = 0.001$$

Sum = 1.000

These are developed knowing the sample size is 3 and the probability of finding a defective from the population of items is 1 in 10 or 0.1 (probability of nondefective is 0.9).

Graph of probability mass function:

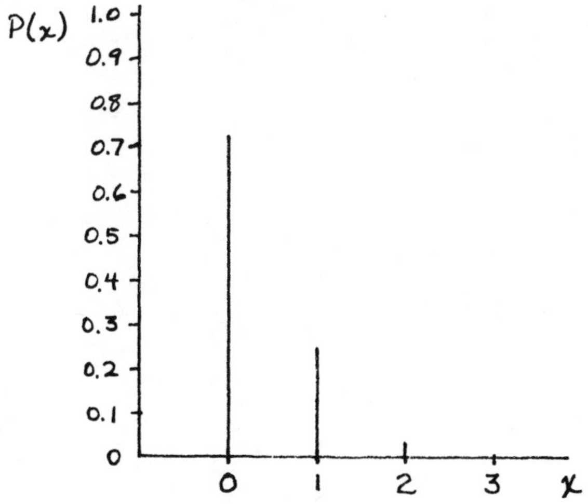

3-10

A "signal" consists of a series of pulses of magnitude X having values 1, 0, -1 each with probability 1/3. A "noise" consists of a series of pulses of magnitude Y having values of 2, 0, -2 with probabilities 1/6, 2/3 and 1/6 respectively. If signal and noise are mixed, the sum consists of pulses of magnitude Z = X + Y. Construct the distribution for Z, and compute its mean & variance.

****

METHOD I

BY A TREE:

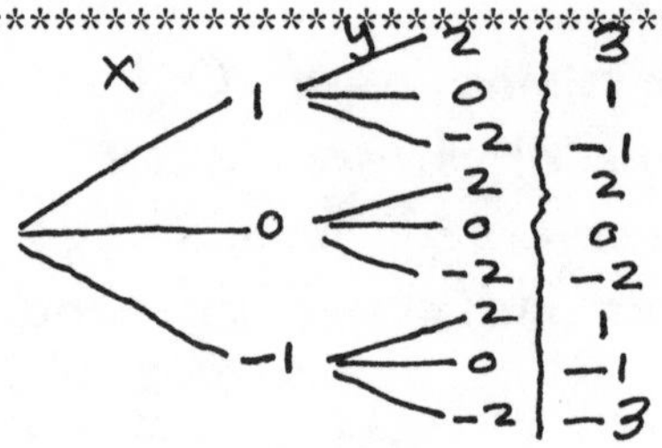

∴ DIST OF Z

Z	p(z)
-3	1/18
-2	1/18
-1	5/18
0	4/18
1	5/18
2	1/18
3	1/18

BY SYMMETRY: $\mu(z) = 0$

$$\sigma^2(z) = E(z^2) = 1 + \frac{4}{9} + \frac{5}{9}$$

$$= 2$$

METHOD II

FOR THE X DISTRIBUTION: $\mu_x = 0,\ \sigma^2_x = 2/3$

FOR THE Y DISTRIBUTION: $\mu_y = 0\ ;\ \sigma^2_y = 4/3$

AS $Z = X + Y$, AND X, Y ARE INDEPENDENT:

$$\mu_Z = \mu_x + \mu_y = 0 + 0 = 0$$

$$\sigma^2_Z = \sigma^2_x + \sigma^2_y = \frac{2}{3} + \frac{4}{3} = 2$$

3-11

A box contains 3 black balls and 1 white ball. A ball is picked and the color is noted. If it's white, the game is over. If it's black, it's discarded and another ball is picked. The game stops when the white ball is picked. Let random variable X equal the number of picks it takes to get the white ball. Find the probability distribution $f_X(x)$ for X.

****

3B 1W

Let r.v. X = Number of picks to get white ball

and event B_i = (Black ball on i th pick)
W_i = (White ball on i th pick)

$$f_X(1) = Pr(X=1) = Pr(W_1) = 1/4$$

$$f_X(2) = Pr(X=2) = Pr(W_2 B_1) = Pr(W_2 | B_1)\, Pr(B_1) = 1/3 \cdot 3/4 = 1/4$$

where W_2B_1 is the compound event consisting of a **black** ball on the 1 st pick followed by a white ball on the 2 nd pick

$$f_X(3) = Pr(X=3) = Pr(W_3 B_2 B_1) = Pr(W_3 | B_2 B_1)\, Pr(B_2 B_1) = Pr(W_3 | B_2 B_1)\, Pr(B_2 | B_1)\, Pr(B_1) = 1/2 \cdot 2/3 \cdot 3/4 = 1/4$$

$$f_X(4) = Pr(X=4) = 1 - \{Pr(X=1) + Pr(X=2) + Pr(X=3)\} = 1 - 3/4 = 1/4$$

Therefore, $f_X(x) = 1/4$, $x = 1, 2, 3, 4$

In other words, it is equally probable that the game will end on the 1 st, 2 nd, 3 rd or 4 th pick.

The use of conditional probabilities has simplified the solution considerably.

3-12

The probability that an electric fuse manufactured by a particular company being defective is known to be 0.1. Since it is costly and time consuming to test all the fuses in a large batch, a quality controller picks 4 fuses at random from the batch and tests each one. Suppose N is the random variable denoting the number of defective fuses in a sample of 4. Determine its probability mass function p(n), and show that

$$\sum_{n=0}^{4} p(n) = 1$$

What is the probability of not more than one fuse tested being defective? What is the probability of two or more fuses that were tested being defective?

****

If one fuse is picked from the batch, the probability of it being defective is 0.1, and it being not defective is $1 - 0.1 = 0.9$.

When a sample of 4 fuses is picked, probability of no defective fuses is

$$P(N=0) = p(0) = 0.9 \times 0.9 \times 0.9 \times 0.9 = (0.9)^4$$

Now suppose only one fuse in the sample of four is bad. There are ${}^4C_1$ different sequences that the bad one could be picked (e.g., the first one picked is bad or the second one picked is bad, and so on). Its probability is

$$P(N=1) = p(1) = {}^4C_1 \times 0.1 \times (0.9)^3$$

Similarly, the probability that only two fuses are defective is,

$$P(N=2) = p(2) = {}^4C_2 \times (0.1)^2 \times (0.9)^2$$

In general, the probability that n fuses are defective

$$P(N=n) = p(n) = {}^4C_n (0.1)^n (0.9)^{4-n}$$

$$n = 0, 1, 2, 3, 4 \quad //$$

$$\sum_{n=0}^{4} P(n) = (0.9)^4 + {}^4C_1 \, 0.1 \times 0.9^3 + {}^4C_2 \, 0.1^2 \times 0.9^2 + {}^4C_3 \, 0.1^3 \times 0.9 + (0.1)^4$$

$$= 0.9^4 + 0.4 \times 0.9^3 + 0.06 \times 0.9^2 + 0.004 \times 0.9 + 0.0001$$

$$= \underline{\underline{1}}$$

Probability of no more than one fuse tested being defective $= P(N \leq 1) = P(0) + P(1)$

$$= 0.9^4 + 0.4 \times 0.9^3 = \underline{\underline{0.9477}}$$

Probability of two or more fuses being defective

$$= P(N > 1) = 1 - P(N \leq 1) = 1 - 0.9477$$

$$= \underline{\underline{0.0523}}$$

■■ **3-13**

A discrete random variable X has been following probability function:

x_i	0	1	2	3	4	5
$p(x_i)$	.05	.10	.20	.40	.15	.10

Find μ and σ^2.

****

$$\mu = \sum_{ALL\,X} x_i \cdot p(x_i) = 0(.05) + 1(.10) + 2(.20) + 3(.40) + 4(.15) + 5(.10)$$

$$= 2.80$$

$$\sigma^2 = \sum_{ALL\,X} x_i^2 \cdot p(x_i) - \mu^2$$

$$= 9.40 - 7.84 = 1.56$$

3-14

A free-lance plumber has collected the following data of his daily occupation [number of times called per day]:

demand	0	1	2	3	4
probability	0.10	0.40	0.25	0.15	0.10

If he charges $40 per call (flat fee) plus parts, find the mean and the variance of his daily fees.

****

LET X = NUMBER OF DAILY CALLS

$$E(x) = \sum_{\text{all } x} x \cdot \text{PROB}(x)$$

$$= (0)(0.10) + (1)(0.40) + (2)(0.25) + (3)(0.15) + (4)(0.10)$$

$$= 1.75$$

$$\sigma^2(x) = E(x^2) - [E(x)]^2 = \sum_{\text{all } x} x^2 \cdot \text{PROB}(x) - 1.75^2$$

$$= (0)(0.10) + (1)(0.40) + (4)(0.25) + (9)(0.15) + (16)(0.10) - 1.75^2$$

$$= 1.2875$$

LET $y = 40x$ FEES CHARGED PER DAY

THEN $E(y) = 40\, E(x)$

$$= (40)(1.75) = 70 \quad [\$]$$

$$\sigma^2(y) = 40^2\, \sigma^2(x)$$

$$= (1600)(1.2875) = 2060 \quad [\$^2]$$

DISTRIBUTION FUNCTION

3-15

A janitorial crew is made up of 10 men and 5 women. The supervisor picks 5 people at random for the night shift. Denoting the number of women assigned to the night shift by the discrete random variable N sketch its probability distribution function F(n). What is the probability of assigning all five women to the night shift?

****

The number of ways (combinations) five people could be picked from the crew $= {}^{15}C_5$

The number of ways five "men" could be picked $= {}^{10}C_5$

∴ Probability of picking no women (hence 5 men) is

$$P(N=0) = P(0) = F(0) = \frac{{}^{10}C_5}{{}^{15}C_5} = 0.0839$$

One woman could be picked from the crew in ${}^5C_1$ ways. For each such selection the four men could be picked in ${}^{10}C_4$ ways. Hence, the probability of picking one woman (hence 4 men) is

$$P(N=1) = P(1) = \frac{{}^5C_1\,{}^{10}C_4}{{}^{15}C_5} = 0.3496$$

similarly

$$P(N=2) = P(2) = \frac{{}^5C_2\,{}^{10}C_3}{{}^{15}C_5} = 0.3996$$

$$P(N=3) = P(3) = \frac{{}^5C_3\,{}^{10}C_2}{{}^{15}C_5} = 0.1499$$

$$P(N=4) = P(4) = \frac{{}^5C_4\,{}^{10}C_1}{{}^{15}C_5} = 0.0167$$

$$P(N=5) = p(5) = \frac{{}^5C_5}{{}^{15}C_5} = 0.003$$

Now by definition, the (cumulative) probability distribution function is given by;

$$F(0) = P(N \leq 0) = p(0) = 0.0839$$

$$F(1) = P(N \leq 1) = F(0) + p(1) = 0.4335$$

$$F(2) = P(N \leq 2) = F(1) + p(2) = 0.8331$$

$$F(3) = P(N \leq 3) = F(2) + p(3) = 0.9830$$

$$F(4) = P(N \leq 4) = F(3) + p(4) = 0.9997$$

$$F(5) = P(N \leq 5) = F(4) + p(5) = 1.0$$

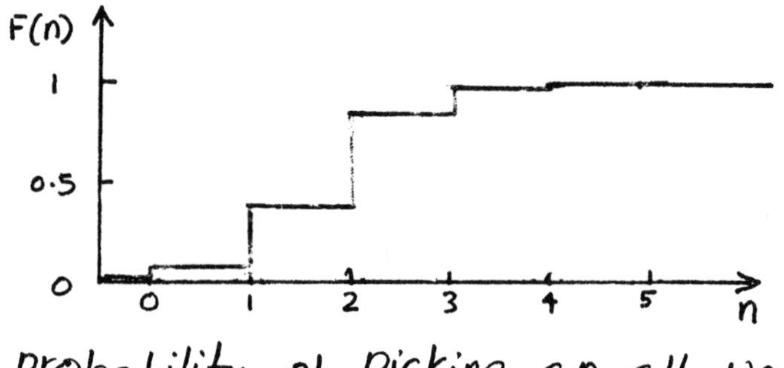

The probability of picking an all-women night crew is

$$p(5) = F(5) - F(4) = \underline{\underline{0.0003}}$$

3-16

You are the president of a communications satellite company and you wish to launch a new communications satellite. You plan to attempt a maximum of three launches, one after the other. You will stop making further launches if you have either a successful launch or if you have three failures in a row. The probability of a successful launch on any one attempt is 0.90, and the outcome of any attempt is independent of the others. Each of the launches you attempt will cost you $50,000,000, but if you are successful in placing a satellite in orbit, your company will earn $100,000,000 (not including the cost of the launches). Of course, if none of the three attempts are successful you have no earnings, but your company still is charged the cost of the attempts.

a) Let N denote the number of trials until a successful launch is made. Find the probability distribution of N.
b) Let P be the random variable representing the total net profit or loss from the entire project. Find the probability distribution for P.
c) Compute the expected value E(P) for the total net profit or loss.

****

S = SUCCESS $\quad Pr\{S\} = 0.9$

N = # OF TRIALS $\quad Pr\{F\} = 1 - Pr\{S\} = 0.1$

a)
$$Pr\{N=1\} = 0.9 = 0.9$$
$$Pr\{N=2\} = 0.1 \times 0.9 = 0.09$$
$$Pr\{N=3\} = 0.1 \times 0.1 \times 0.9 = 0.009$$
$$Pr\{N>3\} = 0.1 \times 0.1 \times 0.1 = 0.001$$
$$\overline{1.000}$$

P = NET PROFIT

b)
$$Pr\{P = \$50M\} = Pr\{N=1\} = 0.9$$
$$Pr\{P = 0\} = Pr\{N=2\} = 0.09$$
$$Pr\{P = -\$50M\} = Pr\{N=3\} = 0.009$$
$$Pr\{P = -\$150M\} = Pr\{N>3\} = 0.001$$
$$\overline{1.000}$$

c)
$$E(P) = \$50M(0.9) + \$0(0.09) - \$50M(0.009) - \$150M(0.001)$$
$$= \$44.4\,M$$

3-17

In order to determine how much cash to stock in a 24-hour automatic bank teller over a weekend, the number of customers arriving to use the teller each 15-minute interval was recorded for 39 intervals. In the order observed, the following numbers of customers arrived each 15 minutes: 3, 4, 5, 10, 3, 5, 2, 6, 3, 10, 4, 5, 4, 2, 7, 3, 3, 8, 7, 9, 3, 1, 2, 2, 4, 5, 10, 6, 3, 1, 8, 1, 3, 3, 2, 11, 3, 3, 1. Draw a cumulative frequency distribution for the number of arrivals per 15-minute period.

****

No. of arrivals k	1	2	3	4	5	6	7	8	9	10	11
No. of 15-minute intervals with k arrivals (frequency)	4	5	11	4	4	2	2	2	1	3	1
Cumulative frequency	4	9	20	24	28	30	32	34	35	38	39

Cumulative frequency distribution:

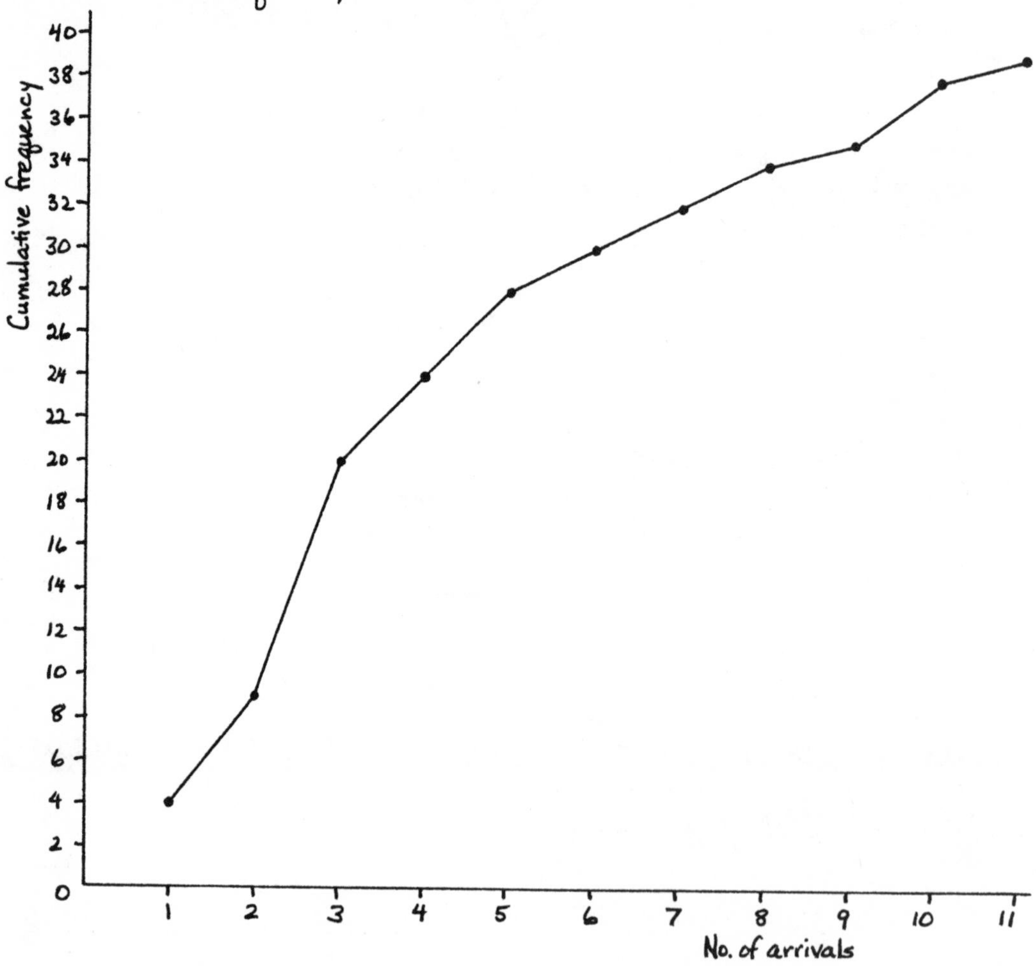

3-18

Given the probability function for a discrete random variable as shown below,

a) find the value of "A".

b) graph the cumulative distribution function.

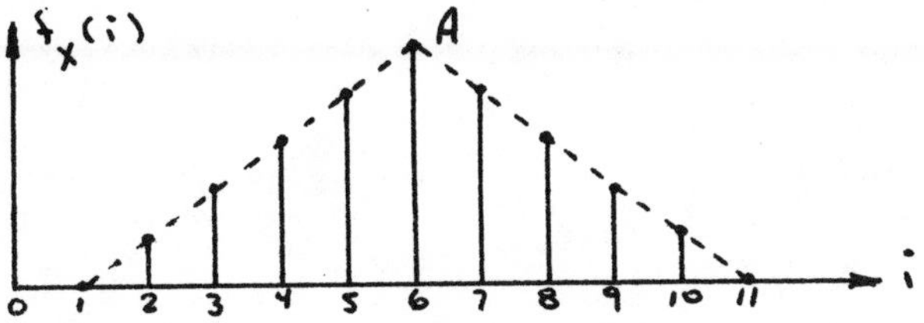

****

a) $$f_X(i) = \begin{cases} A/5\,(i-1) & i = 1,2,3,4,5 \\ A & i = 6 \\ -A/5\,(i-11) & i = 7,8,9,10,11 \end{cases}$$

$$\sum_i f_X(i) = 1 \qquad \sum_{i=1}^{5} \frac{A}{5}(i-1) + A + \sum_{i=7}^{11} -\frac{A}{5}(i-11) = 1$$

By symmetry, it follows that both sums are equal

Hence, $$2\sum_{i=1}^{5} \frac{A}{5}(i-1) + A = 1$$

$$\frac{2A}{5}\left(\sum_{i=1}^{5} i - \sum_{i=1}^{5} 1\right) + A = 1$$

$$\frac{2A}{5}\left[\frac{5(6)}{2} - 5\right] + A = 1 \qquad \text{since } \sum_{i=1}^{n} i = \frac{n(n+1)}{2}$$

$$A = 1/5$$

b) $$F_X(i) = \sum_{k=1}^{i} f_X(k)$$

i	1	2	3	4	5	6	7	8	9	10	11
$F_X(i)$	0	$\frac{1}{25}$	$\frac{3}{25}$	$\frac{6}{25}$	$\frac{10}{25}$	$\frac{15}{25}$	$\frac{19}{25}$	$\frac{22}{25}$	$\frac{24}{25}$	$\frac{25}{25}$	1

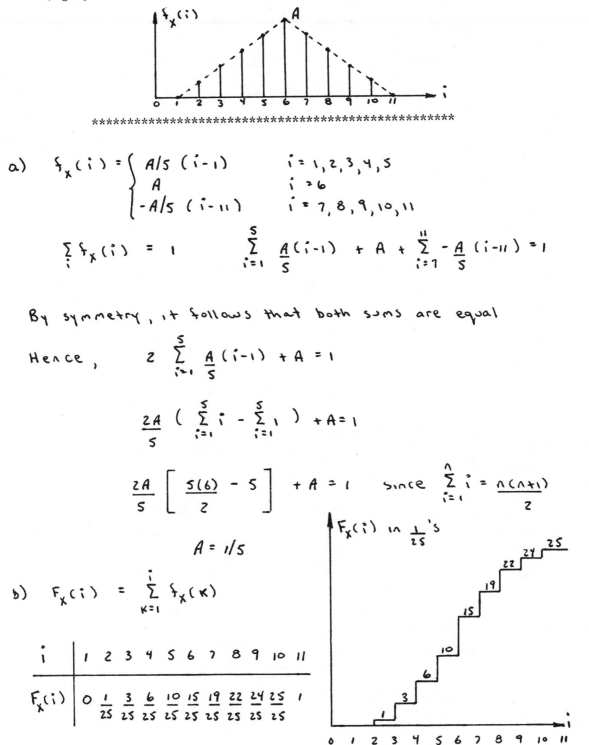

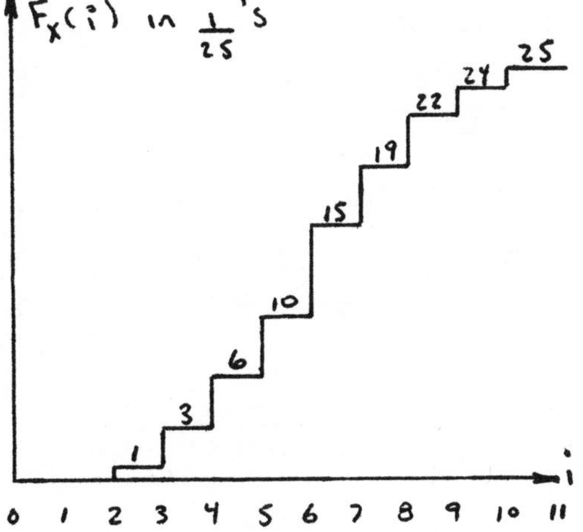

3-19

The discrete random variable X has the following probability distribution.

$$f_X(x) = \frac{A}{x} \qquad x=1,2,3,4,5$$

a) Find the constant A.
b) Find the mean value of X.
c) Sketch the cumulative distribution function, $F_X(x)$.

****

a) $$\sum_{x=1}^{5} f_X(x) = 1 \Longrightarrow \sum_{x=1}^{5} A/x = A\left[1+\frac{1}{2}+\frac{1}{3}+\frac{1}{4}+\frac{1}{5}\right] = 1$$

$$A = \frac{1}{1+\frac{1}{2}+\frac{1}{3}+\frac{1}{4}+\frac{1}{5}} = \frac{60}{137}$$

b) $$\mu_X = \sum_{x=1}^{5} x \cdot f_X(x)$$

$$= \sum_{x=1}^{5} x \cdot \frac{A}{x} = \sum_{x=1}^{5} A = 5A = \frac{300}{137}$$

c) $$F_X(x) = \Pr(X \le x) = \sum_{i=1}^{x} f_X(i)$$

For $x=1$, $F_X(1) = \sum_{i=1}^{1} f_X(i) = f_X(1) = 60/137$

$x=2$, $F_X(2) = \sum_{i=1}^{2} f_X(i) = f_X(1)+f_X(2) = \frac{60}{137}\left[1+\frac{1}{2}\right] = \frac{90}{137}$

$x=3$, $F_X(3) = \sum_{i=1}^{3} f_X(i) = f_X(1)+f_X(2)+f_X(3)$

$$= \frac{60}{137}\left[1+\frac{1}{2}+\frac{1}{3}\right] = \frac{110}{137}$$

In a similar manner,

For $x=4$, $F_X(4) = \frac{125}{137}$

$x=5$, $F_X(5) = 1$

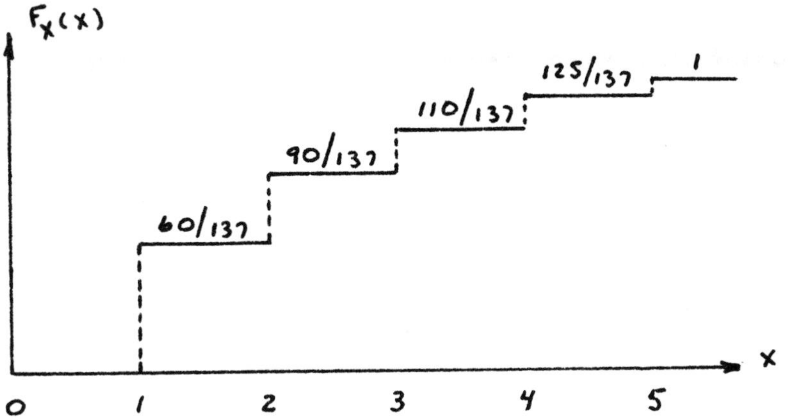

Note, the c.d.f. is a staircase function with jumps at the values of x where the probability distribution is non-zero.

DISCRETE RANDOM VARIABLES - GENERAL

3-20

Suppose that X and Y have the following joint probability function:

Y \ X	2	4
1	0.10	0.15
2	0.20	0.30
3	0.10	0.15

$f_{X,Y}(x,y)$

Find the marginal probability distributions and determine whether X and Y are independent.

****

$$f_X(2) = Pr(X=2) = \sum_y f_{X,Y}(2,y)$$

$$= f_{X,Y}(2,1) + f_{X,Y}(2,2) + f_{X,Y}(2,3)$$

$$= 0.10 + 0.20 + 0.10 = 0.4$$

$$f_X(4) = Pr(X=4) = 1 - Pr(X=2) = 1 - f_X(2) = 0.6$$

$$f_Y(1) = Pr(Y=1) = \sum_x f_{X,Y}(x,1)$$

$$= f_{X,Y}(2,1) + f_{X,Y}(4,1)$$

$$= 0.10 + 0.15 = 0.25$$

$$f_Y(2) = Pr(Y=2) = f_{X,Y}(2,2) + f_{X,Y}(4,2)$$

$$= 0.20 + 0.30 = 0.50$$

$$f_Y(3) = Pr(Y=3) = 1 - Pr(Y=1) - Pr(Y=2) = 0.25$$

Summarizing,

Y \ X	2	4	$f_Y(y)$
1	0.1	0.15	0.25
2	0.2	0.30	0.50
3	0.1	0.15	0.25
$f_X(x)$	0.4	0.6	

X & Y are independent since $f_{X,Y}(x,y) = f_X(x) f_Y(y)$ for $x = 2,4$ & $y = 1,2,3$

e.g. $f_{X,Y}(2,3) = 0.1$

$f_X(2) = 0.4 \quad f_Y(3) = 0.25$

4

CONTINUOUS RANDOM VARIABLES

PROBABILITY DENSITY FUNCTION

4-1

Show that f(x) below is a probability density function and calculate its mean:

$$f(x) = \begin{cases} \dfrac{24}{x^3}, & 3 \leq x \leq 6 \\ 0, & \text{elsewhere} \end{cases}$$

****

First show that f(x) is nonnegative (obvious) and the value of its integral over its nonzero range equals one:

$$\int_3^6 \frac{24}{x^3}\,dx = -\frac{12}{x^2}\Big|_3^6 = -\frac{12}{36} + \frac{12}{9} = \frac{-12+48}{36} = \underline{\underline{1}}$$

Mean = Integral of random variable times probability density function

$$= \int_3^6 x \cdot \frac{24}{x^3}\,dx = \int_3^6 \frac{24}{x^2}\,dx = -\frac{24}{x}\Big|_3^6 = -\frac{24}{6} + \frac{24}{3}$$

$$= -4 + 8 = \underline{\underline{4}}$$

4-2

Let x be a continuous random variable with the following probability density function (pdf).

$$f(x) = x^2 + kx + 1 \qquad \text{for } 0 \leq x \leq 2$$
$$= 0 \qquad \text{elsewhere}$$

Determine the constant k such that f(x) is a valid pdf.

****

The constant k must satisfy two properties:

1. $\int_0^2 f(x)\,dx = 1$

2. $f(x) \geq 0 \quad \text{for} \quad 0 \leq x \leq 2$

Property 1:

$$\int_0^2 (x^2 + kx + 1)\,dx = 1, \quad \left.\frac{x^3}{3} + \frac{kx^2}{2} + x\right]_0^2 = 1$$

$$\frac{8}{3} + 2k + 2 = 1 \quad \Rightarrow \quad k = -\frac{11}{6}$$

Property 2:

By inspection, $x^2 - \frac{11}{6}x + 1 \geq 0$ for $0 \leq x \leq 2$

4-3

Is there a value of k that makes the following function a p.d.f. for some random variable?

$$f_Y(y) = \begin{cases} k - (y-1)^2 & 0 \leq y \leq 2 \\ 0 & \text{otherwise} \end{cases}$$

Hint: It may be helpful to graph the function.

****

For $f_Y(y)$ to be a p.d.f. it must satisfy

$$\int_{-\infty}^{\infty} f_Y(y)\,dy = \int_0^2 \{K - (y-1)^2\}\,dy = 1$$

$$\int_0^2 (K - y^2 + 2y - 1)\,dy = 1$$

$$\left[(K-1)y - \frac{y^3}{3} + y^2\right]_0^2 = 1$$

$$2(K-1) - 8/3 + 4 = 1 \qquad K = 5/6$$

In addition, $f_Y(y) \geq 0$ for all y

Graphing $f_Y(y)$ with $K = 5/6$

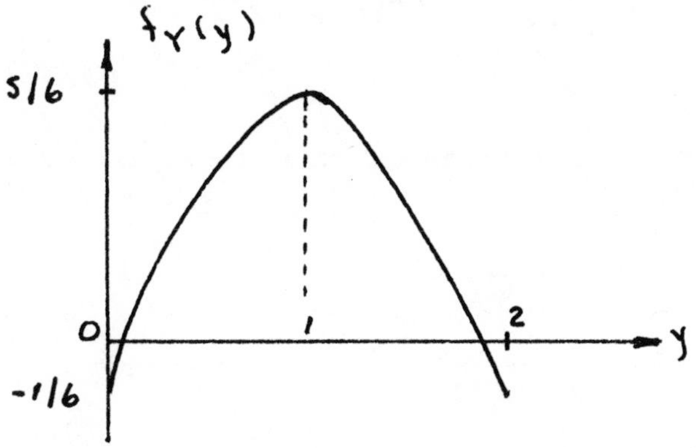

Since $f_Y(y)$ is negative at both ends it cannot be a p.d.f.

4-4

Random variable X has a p.d.f. given by $f_X(x) = \frac{A}{x} \quad 0 \le x \le e^n$; 0 otherwise.

Show that the mean of X, $\mu_X = \frac{1}{n} e^n$.

****

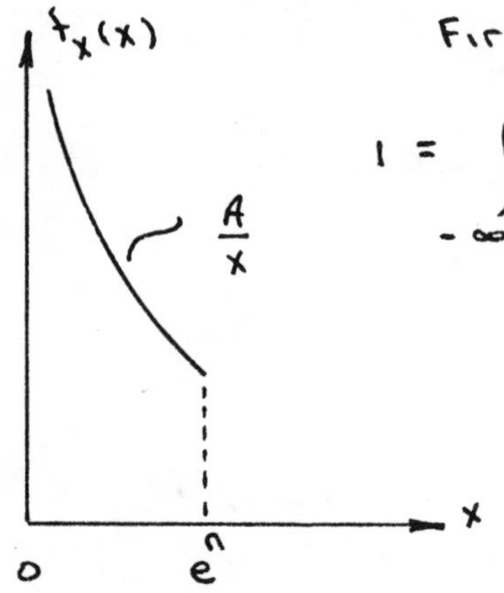

First we need to find "A" from

$$1 = \int_{-\infty}^{\infty} f_X(x)\,dx = \int_0^{e^n} \frac{A}{x}\,dx = A \ln x \Big|_0^{e^n}$$

$$1 = A(\ln e^n - \ln 0)$$

$$1 = An$$

$$A = \frac{1}{n}$$

$$\mu_X = \int_{-\infty}^{\infty} x f_X(x)\,dx = \int_0^{e^n} x \cdot \frac{A}{x}\,dx = A \int_0^{e^n} dx = \frac{1}{n} e^n$$

DISTRIBUTION FUNCTION

4-5

$$\text{Let } P(X \le x) = F(x) = \begin{cases} 0 & \text{if } x < -2 \\ K(x+2) & \text{if } -2 \le x \le 3 \\ 1 & \text{if } x \ge 3 \end{cases}$$

be a distribution function for a random variable X. Find: a) K, b) $P\{1 \le x \le 5/2\}$, c) $P\{x = 1\}$

****

(a) WHEN $X = 3$, $F(x) = 1$ $\therefore$ $5K = 1$ AND $K = \frac{1}{5}$

(b) $P\left(1 \le x \le \frac{5}{2}\right) = F\left(\frac{5}{2}\right) - F(1) = \frac{9}{10} - \frac{3}{5} = \frac{3}{10}$

(c) $P(X = 1) = 0$

4-6

Independent random variables X_1 and X_2 are distributed as follows:

$$f_{X_1}(x_1) = \begin{cases} 1 & 0 \le x_1 \le 1 \\ 0 & \text{otherwise} \end{cases}$$

$$f_{X_2}(x_2) = \begin{cases} 2x_2 & 0 \le x_2 \le 1 \\ 0 & \text{otherwise} \end{cases}$$

Let random variable $Y = \text{Max}(X_1, X_2)$. In other words $Y = \begin{cases} X_1 & \text{if } X_1 \ge X_2 \\ X_2 & \text{if } X_2 \ge X_1 \end{cases}$

Find the c.d.f. for Y, i.e. $F_Y(y)$.

****

$$F_Y(y) = Pr(Y \le y) = Pr[\text{Max}(X_1, X_2) \le y]$$

In order for $\text{Max}(X_1, X_2) \le y$ it follows that both X_1 and X_2 must be less than y

$$F_Y(y) = Pr[X_1 \le y \text{ and } X_2 \le y]$$

X_1 & X_2 Independent, $F_Y(y) = Pr(X_1 \le y)\, Pr(X_2 \le y)$
$= F_{X_1}(y)\, F_{X_2}(y)$

$$F_{X_1}(y) = \int_{-\infty}^{y} f_{X_1}(x_1)\,dx_1 = \begin{cases} 0 & y < 0 \\ y & 0 \le y < 1 \\ 1 & 1 \le y \end{cases}$$

$$F_{X_2}(y) = \int_{-\infty}^{y} f_{X_2}(x_2)\,dx_2 = \begin{cases} 0 & y < 0 \\ y^2 & 0 \le y < 1 \\ 1 & 1 \le y \end{cases}$$

Therefore, $$F_Y(y) = \begin{cases} 0 & y < 0 \\ y^3 & 0 \le y < 1 \\ 1 & 1 \le y \end{cases}$$

Note, the p.d.f. $f_Y(y) = f_{\text{Max}(X_1,X_2)}(y) = \dfrac{d}{dy} F_Y(y)$

$$f_Y(y) = 3y^2 \qquad 0 \le y \le 1$$

4-7

A continuous random variable X has the probability density function $f(x) = 3x^2$; $0 < x < 1$.
a) Find the c.d.f. F(x). b) Find also that $P(x \leq a) = 1/4$. c) Find μ and σ^2 for x.

****

(a) $F(x) = \int f(x)\,dx = \int 3x^2\,dx = x^3 + C$

When $x = 0$, $F(x) = 0$ and $C = 0$

$\therefore F(x) = x^3$

(b) this means $F(a) = \frac{1}{4}$ $\therefore a^3 = \frac{1}{4}$

and $a = \sqrt[3]{\frac{1}{4}} = \frac{\sqrt[3]{2}}{2} = .63$

(c) $\mu = \int_0^1 x \cdot f(x)\,dx = \int_0^1 3x^3\,dx = \left.\frac{3x^4}{4}\right]_0^1 = \frac{3}{4}$

$\sigma^2 = E(x^2) - \mu^2 = \int_0^1 3x^4\,dx - \left(\frac{3}{4}\right)^2$

$= \left.\frac{3x^5}{5}\right]_0^1 - \frac{9}{16} = \frac{3}{5} - \frac{9}{16} = \frac{3}{80} = .0375$

4-8

Is the following function a cumulative distribution function:

$$F(t) = \frac{t^2 + t}{t^2 + 1}, \quad 0 \leq t < \infty ?$$

****

i) $\lim_{t \to 0} F(t) = 0$ ii) $\lim_{t \to \infty} F(t) = 1$

iii) F(t) IS CONTINUOUS

THE ONLY THING LEFT TO CHECK IS WHETHER OR NOT F(t) IS NON-DECREASING, i.e., $f(t) = F'(t)$ MUST BE NON-NEGATIVE.

$$f(t) = \frac{dF}{dt} = \frac{(t^2+1)(2t+1)-(t^2+t)2t}{(t^2+1)^2} = \frac{-t^2+2t+1}{(t^2+1)^2}$$

$f(t) \geq 0$ IF AND ONLY IF $-t^2+2t+1 \geq 0$ OR
$-(t-1-\sqrt{2})(t-1+\sqrt{2}) \geq 0$

CLEARLY, IF $t > 1+\sqrt{2}$, $f(t) < 0$. THUS, $F(t)$ IS DECREASING WHEN $t > 1+\sqrt{2}$, AND CONSEQUENTLY $F(t)$ IS NOT A C.D.F.

JOINT DISTRIBUTIONS

4-9

Consider two independent random variables X and Y with the following probability density functions:

$$f_X(x) = \begin{cases} \frac{1}{5}, & 5 \leq x \leq 10 \\ 0 & \text{otherwise} \end{cases} \quad \text{and } f_Y(y) = \begin{cases} ky^2, & 2 \leq y \leq 4 \\ 0 & \text{otherwise} \end{cases}$$

a) Determine the value of the constant k.

b) Find the joint probability density function of X and Y.

c) Find the conditional probability density function of Y given that X is equal to 8.2.

****

a) $$\int_2^4 f_Y(y)\,dy = 1 \longrightarrow \int_2^4 ky^2\,dy = \frac{ky^3}{3}\Bigg|_2^4 = 1$$

$$k\left[\frac{4^3}{3} - \frac{2^3}{3}\right] = \frac{56k}{3} = 1 \Rightarrow k = \frac{3}{56}$$

b) Since X, Y are independent, $f_{X,Y}(x,y) = f_X(x) \cdot f_Y(y)$

$$f_{XY}(x,y) = \frac{1}{5} \times \frac{3}{56} y^2 = \frac{3}{280} y^2, \text{ for } 5 \leq x \leq 10 \text{ and } 2 \leq y \leq 4$$

c) $$f_{Y|X}(y|8.2) = \frac{f_{XY}(8.2, y)}{f_X(8.2)} = \frac{3y^2}{280} \times 5$$

$$= \frac{3y^2}{56}, \text{ for } 2 \leq y \leq 4$$

4-10

The length and width of a rectangle are jointly distributed according to

$$f_{L,W}(l,w) = \begin{cases} 2e^{-(l+2w)} & l \geq 0 \quad w \geq 0 \\ 0 & \text{otherwise} \end{cases}$$

Find the expected area of the rectangle.

****

$A = LW \qquad E(A) = E(LW)$

$$= \int_{-\infty}^{\infty}\int_{-\infty}^{\infty} lw\, f_{L,W}(l,w)\, dl\, dw$$

$$= \int_0^{\infty}\int_0^{\infty} lw \cdot 2e^{-(l+2w)}\, dl\, dw$$

$$= \int_0^{\infty} 2we^{-2w} \int_0^{\infty} le^{-l}\, dl\, dw$$

$$= \int_0^{\infty} 2we^{-2w}\, dw \cdot \int_0^{\infty} le^{-l}\, dl$$

$$= (1/2)(1) = 1/2$$

Note, both integrals represent the means of exponentially distributed random variables

i.e. $f_X(x) = \lambda e^{-\lambda x} \quad x \geq 0$

$$\mu_x = \int_0^{\infty} x \cdot f_X(x)\, dx = \int_0^{\infty} x \cdot \lambda e^{-\lambda x} = \frac{1}{\lambda}$$

$E(LW)$ reduced to $E(W) \cdot E(L)$ because L and W are independently distributed. This is easily demonstrated by showing that the marginal p.d.f.'s $f_L(l)$ and $f_W(w)$ satisfy

$$f_{L,W}(l,w) = f_L(l) \cdot f_W(w)$$

4-11

An automated continuous strip casting and milling production system has four mills in series in order to properly finish the alloy strip. It is known that each mill has an independent lifetime T_i (i = 1,2,3,4) which follows a Weibull c.d.f. with shape parameter r and scale parameter t_0. Find the c.d.f. for the time to failure of the milling section of the production system, and find the expected time to failure.

****

$$F(t) = Pr\{T_i \le t\} = 1 - e^{-(t/t_0)^r}, \quad 0 \le t < \infty$$

T_S = TIME TO FAILURE FOR SERIES

$$F_S(t) = Pr\{T_S \le t\}$$

$$Pr\{T_S > t\} = Pr\{\min(T_1, T_2, T_3, T_4) > t\}$$

A SERIES FAILS WHEN THE 1ST FAILS

$$= Pr\{T_1 > t, T_2 > t, T_3 > t, T_4 > t\}$$

$$= Pr\{T_1 > t\} Pr\{T_2 > t\} Pr\{T_3 > t\} Pr\{T_4 > t\}$$

INDEPENDENT LIFETIMES

$$= [\exp(-(t/t_0)^r)]^4$$

IDENTICALLY DISTRIBUTED

$$F_S(t) = 1 - e^{-4(t/t_0)^r}$$

$$= 1 - e^{-(t/t_1)^r} \quad \text{WHERE} \quad t_1 = t_0 4^{-1/r}$$

$F_S(t)$ IS A WEIBULL C.D.F. WITH SHAPE PARAMETER = r AND SCALE PARAMETER = t_1.

$$E(T_S) = t_0 4^{-1/r} \Gamma(1 + 1/r)$$

WHERE $\Gamma(\cdot)$ IS THE GAMMA FUNCTION.

4-12

The joint distribution of X_1 and X_2 is given by

$$f_{X_1,X_2}(x_1,x_2) = 3x_1 \quad 0 < x_1 < 1 \quad 0 < x_2 < x_1$$
$$0 \quad \text{otherwise}$$

a) Find the marginal distribution of X_1.
b) Find the conditional density function, $f_{X_2|X_1}(x_2|x_1)$.

****

The joint p.d.f. is non-zero in the triangular region shown below.

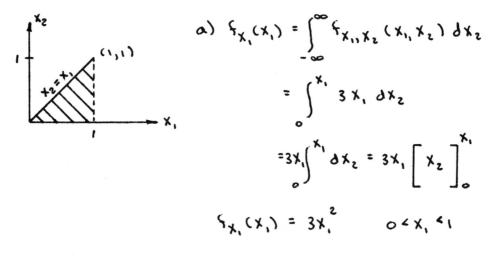

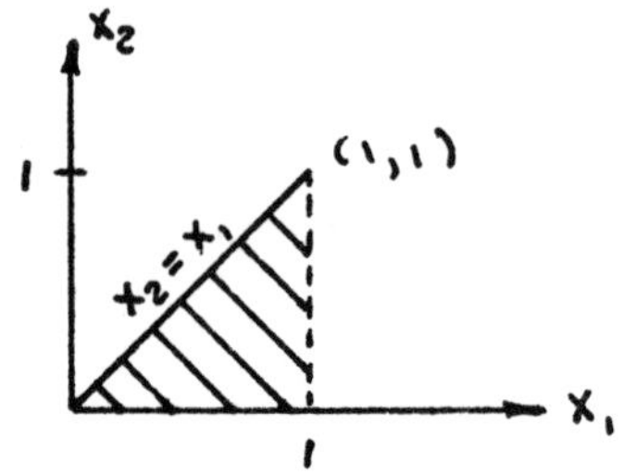

a) $$f_{X_1}(x_1) = \int_{-\infty}^{\infty} f_{X_1,X_2}(x_1,x_2)\,dx_2$$

$$= \int_0^{x_1} 3x_1\,dx_2$$

$$= 3x_1\int_0^{x_1} dx_2 = 3x_1\Big[x_2\Big]_0^{x_1}$$

$$f_{X_1}(x_1) = 3x_1^2 \qquad 0 < x_1 < 1$$

b) $$f_{X_2|X_1}(x_2|x_1) = \frac{f_{X_1,X_2}(x_1,x_2)}{f_{X_1}(x_1)} = \frac{3x_1}{3x_1^2} = \frac{1}{x_1}$$

which applies to the region $0 < x_2 < x_1 < 1$

Note, to check that the above is in fact a probability density function,

$$\int_{-\infty}^{\infty} f_{X_2|X_1}(x_2|x_1)\,dx_2 = \int_0^{x_1} \frac{1}{x_1}\,dx_2 = \frac{1}{x_1}\Big[x_2\Big]_0^{x_1} = 1$$

EXPECTATION

4-13

A young couple have just purchased a new washing machine and are considering buying a warranty policy. The dealer has provided the following historical data on the number of service calls required during the five year warranty period on 50 similar machines.

number of service calls		number of occurrences
0		10
1		22
2		13
3		5
	Total	50

If each service call costs $50, and a five year warranty which will pay for all required service calls costs $125 should they buy the policy?

****

Expected Value $E[V] = \sum_i X_i P(X_i)$

$P(0) = \frac{10}{50} = .2$ $P(1) = \frac{22}{50} = .44$ $P(2) = \frac{13}{50} = .26$ $P(3) = \frac{5}{50} = .1$

E[cost w/o insurance) $= 0(.2) + 50(.44) + 100(.26) + 150(.1) = \$63$

E [cost with ins.) $= \$125$

Do not buy insurance

4-14

A student desires to sell his motorcycle in order to buy a new car. He has been offered $650 as trade in but he feels that he has about a 20% probability of selling it for $750 by just posting notices on the college bulletin board. He also feels that he can increase his probability of selling it for $750 to 75% by advertising in the college newspaper. If the advertisement costs $20 should he advertise?

****

Expected Value $= \sum_i X_i \, P(X_i)$

E[w/o advertise] $= .20(750) + .80(650) = \$730$

E[w advertise] $= .75(750) + .25(650) - 20 = \$705$

Do not advertise

4-15

A circle of random diameter D is inscribed inside a square. Find the expected area between the inside of the square and the outside of the circle.

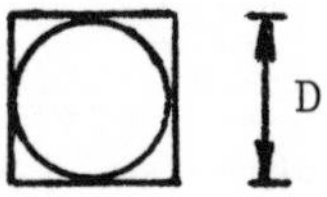

$$f_D(u) = \begin{cases} 1/2 & 1 \leq u \leq 3 \\ 0 & \text{otherwise} \end{cases}$$

****

The area between the inside of the square and the outside of the circle is a random variable given by

$$A = D^2 - \frac{\pi D^2}{4} = \left[1 - \frac{\pi}{4}\right] D^2$$

$$\mu_A = E(A) = E\left\{\left[1 - \frac{\pi}{4}\right] D^2\right\} = \left[1 - \frac{\pi}{4}\right] E(D^2)$$

$$E(D^2) = \int_{-\infty}^{\infty} u^2 f_D(u)\, du = \int_1^3 u^2 \cdot \frac{1}{2}\, du$$

$$= \frac{1}{2}\left[\frac{u^3}{3}\right]_1^3 = \frac{1}{6}(27 - 1) = \frac{13}{3}$$

Therefore, $E(A) = \left[1 - \frac{\pi}{4}\right] \frac{13}{3} = 0.93$

4-16

A continuous random variable X has the probability density function shown below.

$f(x) = 2x^3 - x + 1$ for $0 \le x \le 1$
$f(x) = 0$ elsewhere

(a) Find the cumulative distribution function of X.
(b) Determine the expected value of X.
(c) Using the results from part (a), determine the probability that X is less than or equal to the mean found in part (b).
(d) Comment on the part (c) results.

****

(a) $F(x) = 0$ for $x < 0$
$F(x) = 1$ for $x > 1$

$$F(x) = \int_0^x f(t)\,dt = \int_0^x (2t^3 - t + 1)\,dt = \left.\frac{2t^4}{4} - \frac{t^2}{2} + t\right]_0^x$$

$$F(x) = \frac{x^4}{2} - \frac{x^2}{2} + x \qquad \text{for} \quad 0 \le x \le 1$$

(b) $$E(X) = \int_{-\infty}^{\infty} x\,f(x)\,dx = \int_0^1 x(2x^3 - x + 1)\,dx = \left.\frac{2x^5}{5} - \frac{x^3}{3} + \frac{x^2}{2}\right]_0^1$$

$$= 17/30$$

(c) $$F(17/30) = \frac{(17/30)^4}{2} - \frac{(17/30)^2}{2} + \frac{17}{30} = 0.458$$

(d) Note that, since the p.d.f. is not symmetric, the mean differs from the median.

VARIANCE

4-17

A probability density function is defined as follows:

$$f(x) = \frac{1 + .8x}{2} \quad ; \; -1 < x < 1$$

$$\quad 0 \quad ; \text{ all other } x.$$

a) Show the graph of the density function. b) Find $P(-.5 < x < .5)$.
c) Compute the mean and variance of this density function.

****

(a)

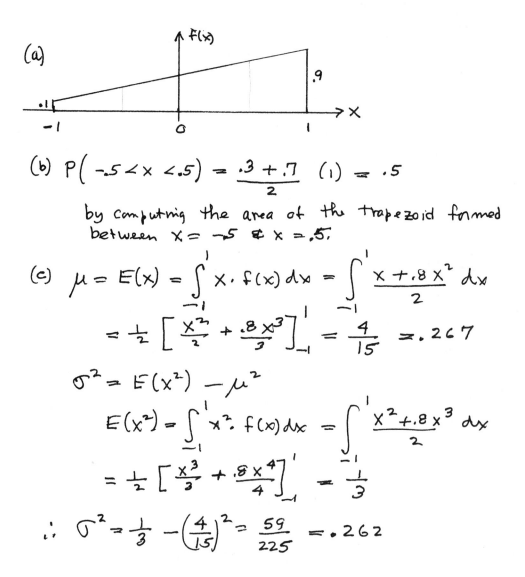

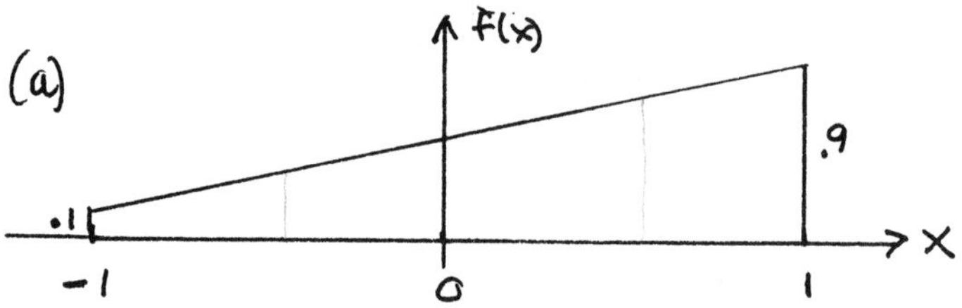

(b) $P(-.5 < x < .5) = \frac{.3 + .7}{2}(1) = .5$

by computing the area of the trapezoid formed between $x = -.5$ & $x = .5$.

(c) $\mu = E(x) = \int_{-1}^{1} x \cdot f(x)\,dx = \int_{-1}^{1} \frac{x + .8x^2}{2}\,dx$

$= \frac{1}{2}\left[\frac{x^2}{2} + \frac{.8x^3}{3}\right]_{-1}^{1} = \frac{4}{15} = .267$

$\sigma^2 = E(x^2) - \mu^2$

$E(x^2) = \int_{-1}^{1} x^2 \cdot f(x)\,dx = \int_{-1}^{1} \frac{x^2 + .8x^3}{2}\,dx$

$= \frac{1}{2}\left[\frac{x^3}{3} + \frac{.8x^4}{4}\right]_{-1}^{1} = \frac{1}{3}$

$\therefore \sigma^2 = \frac{1}{3} - \left(\frac{4}{15}\right)^2 = \frac{59}{225} = .262$

4-18

Let X be a random variable with a p.d.f. of

$$f(x) = 2(2/\pi)^{\frac{1}{2}}\exp(-8(x^2 - 4x + 4)), \quad -\infty < x < \infty .$$

Find the variance of X.

****

f(x) IS VERY CLOSE TO A NORMAL P.D.F. IN FORM.

$$8(x^2 - 4x + 4) = \frac{(x-2)^2}{1/8} = \frac{1}{2}\frac{(x-2)^2}{1/16} = \frac{1}{2}\left[\frac{x-2}{1/4}\right]^2$$

$$2\sqrt{2/\pi} = \sqrt{8/\pi} = 1/\sqrt{2\pi/16}$$

$$f(x) = \frac{1}{\sqrt{2\pi\ 1/16}} \exp\left[-\frac{1}{2}\left(\frac{x-2}{1/4}\right)^2\right]$$

THIS IS IN THE FORM OF A NORMAL P.D.F. THUS

$\mu = 2$ AND $\sigma^2 = 1/16$.

4-19

A test on the life of a sample of pocket calculators produced the results shown below. Calculate the mean life and standard deviation of the calculators.

Life of calculators, hours	90	95	100	105	110	115
Probability	0.10	0.20	0.40	0.15	0.10	0.05

****

$$\mu = \Sigma pX = (0.10)(90) + (0.20)(95) + (0.40)(100) + (0.15)(105) + (0.10)(110) + (0.05)(115)$$

$$= \underline{100.5 \text{ hours}}$$

$$\sigma^2 = \Sigma p(x-\mu)^2 = (0.10)(90-100.5)^2 + (0.20)(95-100.5)^2 + (0.40)(100-100.5)^2 + (0.15)(105-100.5)^2 + (0.10)(110-100.5)^2 + (0.05)(115-100.5)^2$$

$$= 39.75 \text{ (hrs)}^2$$

$$\sigma = \underline{6.3 \text{ hours}}$$

4-20

Random variables X_1 and X_2 are uniformly distributed between 0 and 1. Random variable $Y = X_1 + X_2$ has the p.d.f. shown below. Is it true that $Var(X_1 + X_2) = Var(X_1) + Var(X_2)$?

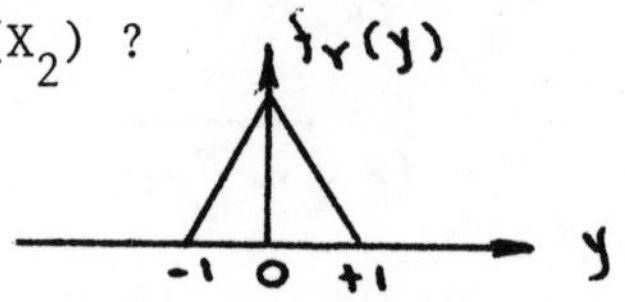

****

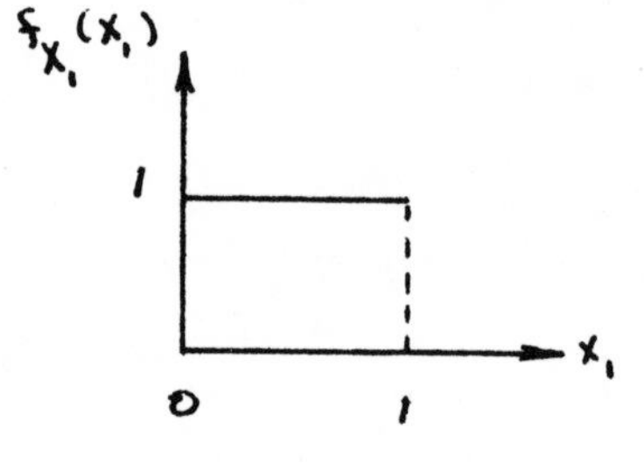

$$Var(X_1) = Var(X_2) = \frac{(1-0)^2}{12} = \frac{1}{12}$$

$$Var(X_1 + X_2) = Var(Y)$$

$$Var(Y) = \int_{-\infty}^{\infty} (y - \mu_Y)^2 f_Y(y)\, dy$$

where $\mu_Y = 0$ since $f_Y(y)$ is symmetric about $y = 0$

$$Var(Y) = \int_{-1}^{1} y^2 f_Y(y)\, dy = 2\int_0^1 y^2 (1-y)\, dy = \frac{1}{6}$$

Therefore, $Var(X_1 + X_2) = \frac{1}{6}$

$$Var(X_1) + Var(X_2) = \frac{1}{12} + \frac{1}{12} = \frac{1}{6}$$

4-21

The mean of the following probability density function is equal to 43/30 (about 1.43). Calculate the variance of the function.

$$f(x) = \begin{cases} \frac{x}{5} & , \quad 0 \le x \le 2 \\ \frac{14.4}{x^4} & , \quad x > 2 \\ 0 & , \quad \text{elsewhere} \end{cases}$$

****

Varianee = Integral of the square of the random variable x times its probability density function over the relevant range <u>minus</u> the square of the mean

$$= \int_0^2 x^2 \cdot \frac{x}{5}\,dx + \int_2^\infty x^2 \cdot \frac{14.4}{x^4}\,dx - \left(\frac{43}{30}\right)^2$$

$$= \int_0^2 \frac{x^3}{5}\,dx + \int_2^\infty \frac{14.4}{x^2}\,dx - \left(\frac{43}{30}\right)^2$$

$$= \left.\frac{x^4}{20}\right|_0^2 - \left.\frac{14.4}{x}\right|_2^\infty - \left(\frac{43}{30}\right)^2$$

$$= \frac{4}{5} - \lim_{x\to\infty} \frac{14.4}{x} + 7.2 - \left(\frac{43}{30}\right)^2$$

$$= 0.8 - 0 + 7.2 - 2.054$$

$$= \underline{\underline{5.946}}$$

CHEBYSHEV'S INEQUALITY

4-22

How large a sample must be taken so that we are 99% certain that the sample mean is within one half of a population standard deviation of the population mean?

****

Let μ = Population mean
σ = Population std. deviation
$\bar{x}$ = Sample mean

Chebyshev's Inequality: $Pr(|\bar{x}-\mu_{\bar{x}}| \le K\sigma_{\bar{x}}) \geq 1 - \frac{1}{K^2}$

where $\mu_{\bar{x}}$ and $\sigma_{\bar{x}}$ are the sampling distribution mean and std. deviation. According to sampling theory,

$$\mu_{\bar{x}} = \mu \quad \& \quad \sigma_{\bar{x}} = \frac{\sigma}{\sqrt{n}}, \qquad n\text{ - sample size}$$

$$Pr\left(|\bar{x}-\mu| \le \frac{K\sigma}{\sqrt{n}}\right) \geq 1 - \frac{1}{K^2}$$

For $\frac{K}{\sqrt{n}} = 0.5 \implies Pr(|\bar{x}-\mu| \le 0.5\sigma) \geq 1 - \frac{1}{(0.5\sqrt{n})^2}$

$$Pr(|\bar{x}-\mu| \le 0.5\sigma) \geq 1 - \frac{4}{n}$$

From the problem statement, we want

$$Pr(|\bar{x}-\mu| \le 0.5\sigma) \geq 0.99$$

Therefore, $1 - \frac{4}{n} = 0.99 \qquad n = 400$

4-23

Random variable X has mean $\mu = 25$ and variance $\sigma^2 = 9$. Find an upper limit on the following probability: $Pr(X \leq 16 \text{ or } X \geq 34)$

****

$$Pr(X \leq 16 \text{ or } X \geq 34) = Pr(|X-25| \geq 9)$$

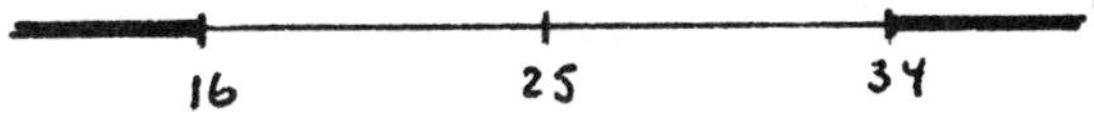

Chebyshev's Inequality states

$$Pr(|X-\mu| \geq K\sigma) \leq \frac{1}{K^2}$$

Solving for K,

$$K\sigma = 9, \quad K = \frac{9}{\sigma} = \frac{9}{3} = 3$$

Consequently, $Pr(|X-25| \geq 9) = Pr(|X-\mu| \geq 3\sigma)$

$$\leq \frac{1}{9}$$

4-24

Determine the probability that a random variable will fall within 2.5 standard deviations of its mean:

a) when the distribution is unknown;

and, b) assuming normality.

****

a) FROM CHEBYSHEV:

$$P(\mu - m\sigma < X < \mu + m\sigma) \geq 1 - \frac{1}{m^2}$$

$$\geq 1 - \frac{1}{2.5^2} = 84\%$$

b) FROM A NORMAL TABLE:

$$P(\mu - m\sigma < X < \mu + m\sigma)$$

$$= P(-m < Z < m) = 98.76\%$$

4-25

In a popular restaurant, the waiting time to get a table during the lunch hour may be represented by the random variable T with mean value μ = 10 minutes and standard deviation σ = 1.5 minutes. On a particular day, a customer had to wait 25 minutes before she could get a table. Could this be an unusual experience at this restaurant? How would your conclusion change if σ = 15 minutes?

****

Use Chebyshev's inequality;

$$P(|T-\mu| \geq r\sigma) \leq \frac{1}{r^2} \quad \text{for positive } r.$$

For a waiting time of 25 min, deviation from the mean value $= 25 - 10 = 15$

Setting $r\sigma = 15$ we get $r = \frac{15}{\sigma} = \frac{15}{1.5} = 10$

Hence from Chebyshev's inequality

$$P(|T-10| \geq 15) \leq \frac{1}{10^2} = 0.01$$

It follows that a deviation of 15 min or more is an unusual (small probability) occurrence.

If $\sigma = 15$ $\quad r = \frac{15}{15} = 1$

Hence $P(|T-10| \geq 15) \leq \frac{1}{1^2} = 1$

Then a deviation of 15 min or more becomes a very common occurrence!

CENTRAL LIMIT THEOREM

4-26

The time to serve a customer at a store is randomly distributed according to

$$f_T(t) = \frac{1}{5} e^{-t/5} \qquad t \geq 0$$

where t is measured in minutes. Find the probability that a line of 30 customers waiting to be served can be completely served in less than 3 hours.

****

Let T_i = Time to serve the i th customer

$$Pr\left(\sum_{i=1}^{30} T_i \leq 180\right) = Pr\left(1/30 \sum_{i=1}^{30} T_i \leq 6\right)$$

$$= Pr(\bar{T} \leq 6)$$

Applying the Central Limit Theorem and converting $\bar{T}$ to the standard Normal random variable

$$= Pr\left[\frac{\bar{T} - \mu_{\bar{T}}}{\sigma_{\bar{T}}} \leq \frac{6 - \mu_{\bar{T}}}{\sigma_{\bar{T}}}\right]$$

where $\mu_{\bar{T}} = \mu_T = 5$

$$\sigma_{\bar{T}} = \frac{\sigma_T}{\sqrt{30}} = \frac{5}{\sqrt{30}}$$

$$= Pr\left[Z \leq \frac{6-5}{5/\sqrt{30}}\right]$$

$$= Pr(Z \leq 1.0954)$$

$$= 0.8633$$

4-27

A hardware store packages washers in bags of forty. If washers have a mean weight of ¼ ounce and a standard deviation of 1/10 ounce, what is the chance that a bag will weigh more than eleven ounces?

****

AS $n > 30$, ASSUME NORMALITY (CENTRAL LIMIT THEOREM)

$$\mu_{BAG} = 40\left(1/4\right) = 10, \quad \sigma_{BAG} = \sqrt{40}\left(1/10\right) = .6325$$

$$P(X > 11) = P\left(Z > \frac{11-10}{.6325}\right) = P(Z > 1.5811)$$

$$= 5.69\,\%$$

4-28

A research firm is interested in determining the average cost of a hospital confinement in a certain area. They wish to estimate the true mean cost within $10 with a 95% confidence. They make a pilot study of 50 patients and determine that the mean is $1,872 with a standard deviation of $100. How many additional patients must be surveyed to obtain the desired accuracy?

****

$$n = \left(\frac{Z\sigma}{\bar{X}-\mu}\right)^2 = \left[\frac{(1.96)(100)}{10}\right]^2 = 384.16$$

$$n \simeq 385$$

Since 50 patients have previously been sampled 335 additional patients will be required.

4-29

The range values of ten samples, each of size 5, from a population are given as follows: 8, 3, 5, 8, 6, 3, 3, 4, 8, 7. The population mean is equal to 48. What would be the risk of a sample mean (from a sample of size 5) greater than 50?

****

According to the Central Limit Theorem, the distribution of sample means is normal with mean 48 and standard deviation

$s_{\bar{x}} \cong \frac{s}{\sqrt{n}}$, where s is the population standard deviation estimate and n is the sample size of 5.

s is approximated by $\bar{R} \cdot d$ ($\bar{R}/d_2$ in some texts), where $\bar{R}$ represents the average range of the 10 samples above and d is a range coefficient obtained from a table for the appropriate sample size. Thus,

$$s \cong \bar{R} \cdot d = \frac{8+3+5+8+6+3+3+4+8+7}{10} \cdot (0.4299) = 2.36 .$$

Then $s_{\bar{x}} \cong \frac{2.36}{\sqrt{5}} = 1.06$

The risk (probability) of a sample mean greater than 50 is shown on the normal distribution of sample means to the right.

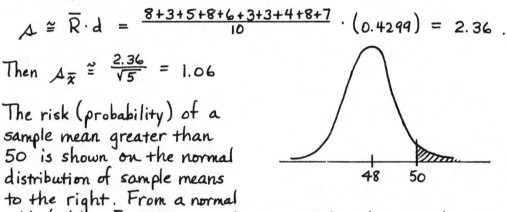

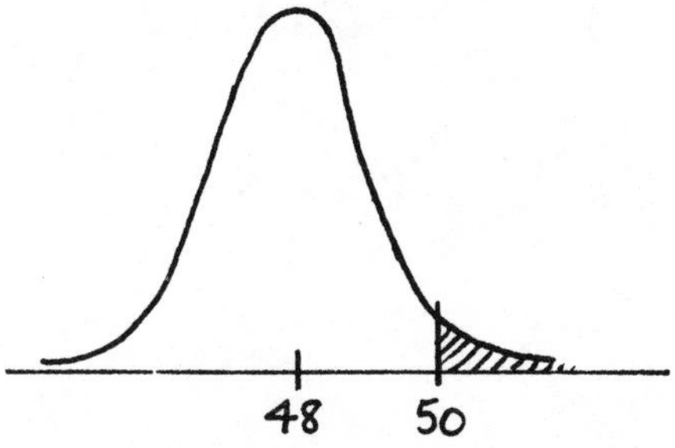

From a normal table (where Φ represents the area under the normal curve from $-\infty$ to the desired value), the desired probability is

$$1 - \Phi\left(\frac{50-48}{1.06}\right) = 1 - \Phi(1.89) = 1 - 0.9706 = 0.0294$$

or about a <u>2.94%</u> risk.

4-30

A box contains 150 screws where the expected weight of an individual screw is one ounce. A crate is packaged with 10 boxes for transportation purposes from the manufacturing plant to several distribution centers. If the variance per box is 2.5 ounces^2, determine the probability that a crate selected at random weighs more than 1,510 ounces.

****

This is a case of a normal distribution as dictated by central limit theorem.

Expected weight of a crate =

$$\mu = 150 \times 1 \times 10 = 1500 \text{ ounces}$$

variance of a crate $= \sigma^2 = 2.5 \times 10 = 25$ ounces^2

If P = probability, and X = weight of a crate picked at random, then

$$P(X > 1510) = 1 - P(X \leq 1510)$$

$$= 1 - P\left(\frac{X-\mu}{\sigma} \leq \frac{1510-1500}{\sqrt{25}}\right)$$

$$= 1 - P(Z \leq 2) = 1 - \Phi(2)$$

Where Φ is a standard normal distribution function. Selecting its value from appropriate probability tables, one has

$$P(X > 1510) = 1 - \Phi(2) = 1 - 0.9772 = 0.0228$$
$$\simeq 2.3\,\%$$

4-31

Salaries of salesmen for a large company are Uniformly distributed between ten and twenty thousand dollars per year. Find the probability that the total salaries of 50 salesmen chosen at random exceeds seven hundred and sixty thousand dollars.

****

Let random variable X = Salary of randomly chosen salesman

The p.d.f. of X is $f_X(x) = \begin{cases} 1/10 & 10 \le x \le 20 \\ 0 & \text{otherwise} \end{cases}$

$$\Pr\left(\sum_{i=1}^{50} X_i > 760\right) = \Pr\left(1/50 \sum_{i=1}^{50} X_i > 1/50 \cdot 760\right)$$

where X_i, $i = 1, 2, 3, \ldots\ldots, 50$ is the random sample of 50 salaries

$$= \Pr(\bar{X} > 15.2)$$
$$= 1 - \Pr(\bar{X} \le 15.2)$$

$$= 1 - \Pr\left[\frac{\bar{X} - \mu_X}{\sigma_X/\sqrt{n}} \le \frac{15.2 - \mu_X}{\sigma_X/\sqrt{n}}\right]$$

$$= 1 - \Pr\left[Z \le \frac{15.2 - 15}{(5/\sqrt{3})/\sqrt{50}}\right]$$

where $\mu_X = \frac{10+20}{2}$ & $\sigma_X = \frac{20-10}{\sqrt{12}}$

i.e. mean and std. deviation of $f_X(x)$

$$= 1 - \Pr(Z \le 0.4899)$$
$$= 1 - 0.6879$$
$$= 0.312$$

Note, the Central Limit Theorem was used to assume $\bar{X}$ is approximately Normally distributed.

5
DISCRETE DISTRIBUTIONS

UNIFORM

5-1

When an honest die (one half of a pair of dice) is rolled, the result is one of the set, {1,2,3,4,5,6}, with each result being equally likely. What is the probability that the result of a roll of a pair of dice will be:

a) 12?
b) 3?
c) 7?

****

a) $P(12) = P[6 \text{ ON 1ST DIE AND } 6 \text{ ON 2ND DIE}]$

$= (1/6)(1/6) = 1/36$

b) $P(3) = P[(2 \text{ AND } 1) \text{ OR } (1 \text{ AND } 2)]$

$= 1/36 + 1/36 = 1/18$

c) $P(7) = P[(4 \text{ AND } 3) \text{ OR } (3 \text{ AND } 4)]$

$+ P[(5 \text{ AND } 2) \text{ OR } (2 \text{ AND } 5)]$

$+ P[(6 \text{ AND } 1) \text{ OR } (1 \text{ AND } 6)]$

$= 6/36 = 1/6$

5-2

Discrete random variable X has a probability function given by

$$f_X(x) = \begin{cases} \frac{1}{n} & x = k+1,\ k+2,\ \ldots\ldots,\ k+n \\ 0 & \text{otherwise} \end{cases}$$

Show that the mean of X is the average of k+1 and k+n.

****

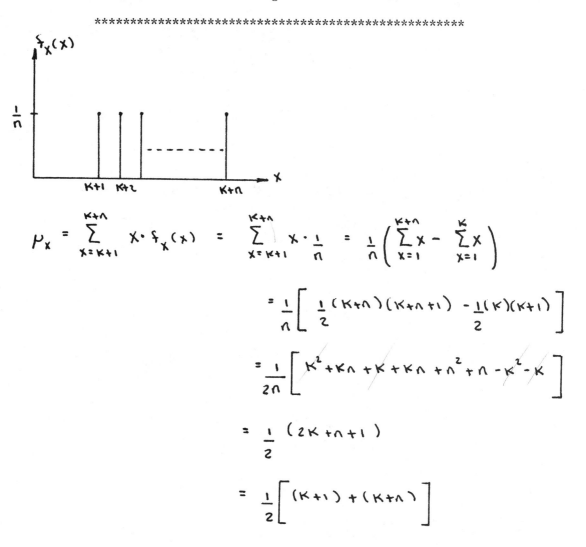

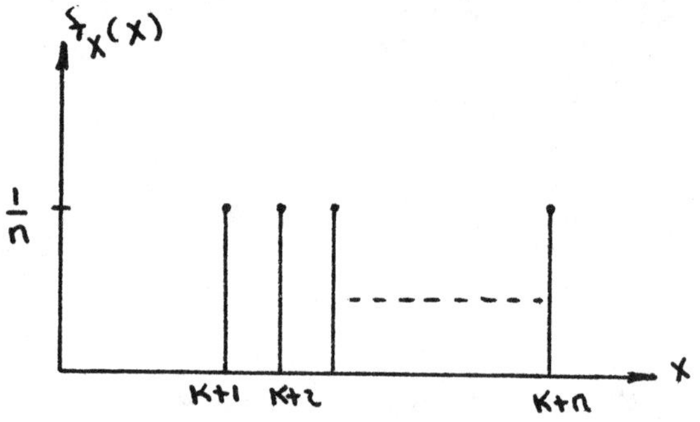

$$\mu_X = \sum_{x=k+1}^{k+n} x \cdot f_X(x) = \sum_{x=k+1}^{k+n} x \cdot \frac{1}{n} = \frac{1}{n}\left(\sum_{x=1}^{k+n} x - \sum_{x=1}^{k} x\right)$$

$$= \frac{1}{n}\left[\frac{1}{2}(k+n)(k+n+1) - \frac{1}{2}(k)(k+1)\right]$$

$$= \frac{1}{2n}\left[k^2 + kn + k + kn + n^2 + n - k^2 - k\right]$$

$$= \frac{1}{2}(2k+n+1)$$

$$= \frac{1}{2}\left[(k+1) + (k+n)\right]$$

5-3

Consider the discrete random distribution P(x)=1/(n+1), where x=0,1,...,n.

(a) Determine the moment generating function $M(t)=E(e^{tx})$.

(b) Show that M(0)=1

(c) Show that M'(0)=n/2

(d) Show that $M''(0)=\dfrac{2n^3+3n^2+n}{6(n+1)}=V(x)+n^2/4$

where:

$$\sum_{x=1}^{n} x = n(n+1)/2 \quad \text{and} \quad \sum_{x=1}^{n} x^2 = (2n^3+3n^2+n)/6$$

****

(a) $$M(t)=\sum_{x=0}^{n}\frac{1}{n+1}e^{xt} = \frac{1}{n+1}\left[1+e^t+e^{2t}+\cdots+e^{nt}\right]$$

$$M(0)=\frac{1}{n+1}\left[1+1+1+\cdots+1\right] = \frac{n+1}{n+1}=1$$

(b) $$M'(t)=\frac{1}{n+1}\left[e^t+2e^{2t}+\cdots+ne^{nt}\right]$$

$$M'(0)=\frac{1}{n+1}\left[1+2+\cdots+n\right]$$

BUT $$\sum_{x=1}^{n} x = \left[1+2+\cdots+n\right] = \frac{n(n+1)}{2}$$

$$M'(0)=\frac{1}{n+1}\left[\frac{n(n+1)}{2}\right] = n/2$$

(c) $$M''(t)=\frac{1}{n+1}\left[e^t+(2)^2e^{2t}+\cdots+n^2e^{nt}\right]$$

$$M''(0)=\frac{1}{n+1}\left[1+(2)^2+\cdots+n^2\right]$$

BUT $$\sum_{x=1}^{n} x^2 = (2n^3+3n^2+n)/6$$

$$\therefore M''(0)=\frac{2n^3+3n^2+n}{6(n+1)} = V(x)+\left[E(x)\right]^2$$

$$= V(x)+n^2/4$$

5-4

If someone were to give us $10 each time we roll a "1" or a "6" with a fair die, how much should we pay them when we roll a "2", "3", "4" or "5" to make the game equitable?

****

Let r.v. X = Number appearing on die

The probability distribution of X is $f_X(x) = 1/6 \quad x = 1, 2, \ldots, 6$

Let A = Payment when rolling 2, 3, 4, 5

Expected Winnings = $\$10 \cdot Pr(X = 1, 6) + A \cdot Pr(X = 2, 3, 4, 5)$

$$Pr(X = 1, 6) = Pr(X = 1) + Pr(X = 6)$$

$$= f_X(1) + f_X(6) = 2/6$$

$$Pr(X = 2, 3, 4, 5) = Pr(X = 2) + Pr(X = 3) + Pr(X = 4) + Pr(X = 5)$$

$$= f_X(2) + f_X(3) + f_X(4) + f_X(5) = 4/6$$

To make the game equitable, the expected winnings must be zero.

i.e. $10(2/6) + A(4/6) = 0$

$A = -5$ Ans. $5

BINOMIAL

5-5

A certain product is received in lots of 500 units. The quality control acceptance procedure is to choose 20 units at random from the lot and test them. If 3 or more units in the sample are found defective the entire lots is rejected. What is the probability that a lot containing 40 defective parts will be accepted?

****

Since sampling is without replacement the distribution is hypergeometric

$N = 500 \qquad G = 460$

$n = 20 \qquad D = 40$

However, since $N > 20n$ the Binomial is a valid approximation to the hypergeometric. Lot will be accepted if the sample contains 0, 1 or 2 defects

$p = \frac{40}{500} = .08 \qquad q = \frac{460}{500} = .92 \qquad n = 20$

$P(x) = C^n_x \, p^x \, q^{n-x}$

$P(0) = C^{20}_0 (.08)^0 (.92)^{20} = 1.887$

$P(1) = C^{20}_1 (.08)^1 (.92)^{14} = .3281$

$P(2) = C^{20}_2 (.08)^2 (.92)^{18} = \underline{.2711}$

Probability of acceptance = .7879

5-6

During a certain medical campaign a random sample of 7 people was drawn from a large population of whom 70% are smokers. What is the probability that exactly 4 people in the sample will be smokers?

****

This is a case of a binomial distribution.

The number of people in the random sample $= n = 7$

The number of exact smokers in the sample $= k = 4$

The probability of smokers in the population $= p = 0.7$

The probability of nonsmokers in the population $= 1 - p = 0.3$

The probability of exact number of smokers in the sample $= p(k)$, or

$$p(k) = \binom{n}{k} p^k (1-p)^{n-k}$$

or, $p(k) = \binom{7}{4}(0.7)^4(0.3)^3 = \frac{7!}{4!\,3!}(0.7)^4(0.3)^3$

or, $p(k) = 0.2269$, or, $p(k) = 23\%$

5-7

We are told that the average salary of starting engineers is \$20,000 per year, with a standard deviation of \$2,200.

a) A placement firm, interested in more information about recent engineering graduates, conducts a phone survey of 40 randomly selected starting engineers. Calculate the probability that the average salary for this sample is within ± \$800 of the population mean.

b) The placement firm will ultimately obtain five different and independently selected samples, each consisting of 40 starting engineers. Calculate the probability that 3 out of the 5 samples will have a sample mean within ± \$800 of the population mean. Make sure that all your assumptions are clearly stated.

****

a) Let $\bar{X}$ denote the average salary of a sample of 40 starting engineers.

$$\bar{X} \underset{\text{approx.}}{\sim} N\left(20{,}000, \frac{(2{,}200)^2}{40}\right) \qquad \text{[by CLT]}$$

$$\text{proba}(19{,}200 \leq \bar{X} \leq 20{,}800)$$

$$= \text{proba}\left(\frac{19{,}200 - 20{,}000}{2{,}200/\sqrt{40}} \leq Z \leq \frac{20{,}800 - 20{,}000}{2{,}200/\sqrt{40}}\right)$$

$$= \text{proba}(-2.30 \leq Z \leq +2.30) = \underline{\underline{0.9786}}$$

b) Each sample can be viewed as an independent Bernoulli trial, where SUCCESS $\Leftrightarrow$ avg. salary for sample is within ± \$800.

Proba(success) = 0.9786 [from part a] — constant across trials.

We are interested in the random variable:

Y = # of successes in 5 independent Bernoulli trials.

Under these assumptions, Y is Binomially distributed, with $n = 5$ and $p = 0.9786$.

$$\therefore \text{proba}(Y=3) = \binom{5}{3}(0.9786)^3(1-0.9786)^2$$

$$= \underline{\underline{0.0043}}.$$

5-8

Among a shipment of 2000 cans of beer, 400 were sent with less volume than they should. If one purchases at random ten of these cans,

(a) What's the probability that exactly 2 will have less beer than they should?

(b) What's the probability that at least 8 cans will have the correct volume content?

****

SITUATION IS HYPERGEOMETRIC WITH

$N = 2{,}000 \qquad a = 400 \qquad n = 10$

a) $\Pr(X = 2;\ N = 2{,}000,\ a = 400,\ n = 10)$

$$= \frac{\binom{400}{2}\binom{1600}{8}}{\binom{2000}{10}}$$

Using binomial approximation:

$$\Pr(X = 2) \simeq b(2;\ 10,\ p = 0.20)$$

$$= \binom{10}{2} 0.20^2\ 0.80^8 = 0.3020$$

b) Pr (at least 8 with correct volume)

= Pr (at most 2 with less volume than they should)

$$= \Pr(X \le 2) = h(0) + h(1) + h(2)$$

$$\simeq B(2;\ 10,\ 0.20)$$

$$= 0.6778$$

5-9

15 castings of a certain type are produced per day in a foundry. The finished castings are inspected and classified as defective or nondefective. Records indicate that of the last 500 castings inspected, 16 were defective. Based on this information, find the probability of having at least two defective castings in a day's production.

****

Assuming occurrences of defective castings are independent, and classifying an inspected casting into one of two categories (defective or nondefective), the binomial distribution will provide the most accurate value of the desired probability.*

First find the process average fraction defective for castings

$$p = \frac{16}{500} = 0.032$$

Then the probability of at least 2 defective castings in a day's production of 15 castings is a cumulative value of the binomial distribution:

$$P(\text{at least 2 defectives in 15}) = P(\text{2 in 15 or 3 in 15 or 4 in 15 or} \ldots \text{15 in 15})$$

$$= \binom{15}{2}(0.032)^2(1-0.032)^{13} + \binom{15}{3}(0.032)^3(1-0.032)^{12} + \cdots + \binom{15}{15}(0.032)^{15}(1-0.032)^0$$

This is difficult computationally but equivalent to

$$1 - P(\text{less than 2 in 15 defectives})$$

$$= 1 - P(\text{0 in 15 or 1 in 15})$$

$$= 1 - \left[\binom{15}{0}(0.032)^0(1-0.032)^{15} + \binom{15}{1}(0.032)^1(1-0.032)^{14}\right]$$

$$= 1 - 0.614 - 0.304$$

$$= \underline{\underline{0.082}}$$

* Note that the Poisson distribution will provide a fairly good approximation in this case (0.084) since p is quite small.

5-10

Let x be a binomially distributed random variable with mean E(x)=4, and variance V(x)=4/3.

(a) Find the distribution of x, given that $P(x)=(n!/((n-x)!x!))p^{x}(1-p)^{n-x}$.

(b) From (a) show that E(x)=4.

(c) From parts (a) and (b) show that V(x)=4/3.

****

(a) $E(x) = np = 4, \quad V(x) = np(1-p) = 4/3$

$\therefore \; 4(1-p) = 4/3, \quad 1-p = 1/3 \quad \therefore \; p = 2/3$

$n = 6$

$$P(x) = \binom{n}{x} p^{x}(1-p)^{n-x} = \binom{6}{x}\left(\frac{2}{3}\right)^{x}\left(\frac{1}{3}\right)^{6-x}$$

(b) $$E(x) = \sum_{x=0}^{n} x \frac{n!}{(n-x)!\,x!} p^{x}(1-p)^{n-x}$$

$$E(x) = \sum_{x=0}^{6} x \frac{6!}{(6-x)!\,x!} (2/3)^{x}(1/3)^{6-x}$$

OR $x=1$

$$E(x) = \sum_{x=1}^{6} \frac{6!}{(6-x)!\,(x-1)!} (2/3)^{x}(1/3)^{6-x}$$

$$= 6\left(\frac{2}{3}\right)\sum_{x=1}^{5} \frac{5!}{(5-(x-1))!\,(x-1)!}\left(\frac{2}{3}\right)^{x-1}(1/3)^{5-(x-1)} = 4$$

(c) $$V(x) = E(x^2) - \{E(x)\}^2 = \sum_{x=0}^{6} x^2 \binom{6}{x}(2/3)^{x}(1/3)^{6-x} - (4)^2$$

$$V(x) = \sum_{x=2}^{6} x(x-1)\frac{6!}{(6-x)!\,x!}(2/3)^{x}(1/3)^{6-x}$$

$$+ \sum_{x=1}^{6} x\binom{6}{x}(2/3)^{x}(1/3)^{6-x} - 16$$

$$V(x) = \sum_{x=2}^{6} \frac{6!}{(4-(x-2))!\,(x-2)!} (2/3)^{x} (1/3)^{4-(x-2)} + 4 - 16$$

$$= 6(5)(2/3)^2 \sum_{x=2}^{4} \binom{4}{x-2} (2/3)^{x-2} (1/3)^{4-(x-2)} - 12$$

$$= 30(4/9) - 12 = 40/3 - 12 = \frac{40-36}{3}$$

$$\therefore V(x) = 4/3$$

5-11

The traffic light at Main St. and Broadway is either green, red or yellow for Main St. traffic with the following probabilities:

Pr(green) = 0.7 Pr(red) = 0.25 Pr(yellow) = 0.05

Find the probability that at least 3 out of 5 cars on Main St. get a green light at the intersection.

****

Let r.v. X = Number of cars out of five that get a green light

Then X is binomially distributed, $b(n, p)$ where $p = Pr(green) = 0.7$ & $n = 5$

$$Pr(X \geq 3) = 1 - Pr(X < 3)$$
$$= 1 - Pr(X \leq 2)$$
$$= 1 - B(2; 5, 0.7)$$
$$= 1 - [1 - B(2; 5, 0.3)]$$
$$= B(2; 5, 0.3)$$
$$= 0.8369$$

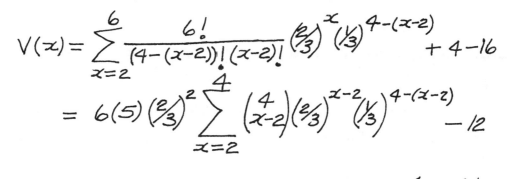

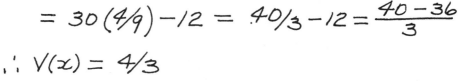

5-12

An experimental drug is claimed to be 90% effective in curing a certain disease. If the claim is correct

a) How many out of 20 people with the disease are expected to recover after being given the drug?

b) Find the probability that less than 18 out of 20 people will be cured.

****

Let r.v. X = Number of people cured out of 20

Then $X \sim b(n, p)$ with $n = 20$ & $p = 0.9$

a) Expected number to recover $= \mu_X = np = 20(0.9) = 18$

b) $Pr(X < 18) = Pr(X \leq 17)$

$$= B(17; 20, 0.9)$$
$$= 1 - B(2; 20, 0.1)$$
$$= 1 - 0.6769$$
$$= 0.3231$$

Note, since tables of the binomial distribution stop at $p = 0.5$, use of the following relationship was necessary,

$$B(x; n, p) = 1 - B(x-1; n, 1-p)$$

5-13

If the probability of hitting a target is .2 and ten shots are fired independently, what is the probability that the target will be hit at least once? At least twice?

****

(a) $P(X \geq 1) = 1 - P(X = 0) = 1 - (.8)^{10} = 1 - .1074$

$$= .8926$$

(b) $P(X \geq 2) = 1 - P(X = 0) - P(X = 1)$

$$= 1 - (.8)^{10} - 10(.8)^9 \cdot (.2) = .6242$$

5-14

THE PROBABILITY THAT A FUSE WILL BE DEFECTIVE WHEN FIRST INSTALLED IS 0.08.

IF SIX FUSES ARE SELECTED AT RANDOM., WHAT IS THE PROBABILITY THAT LESS THAN TWO FUSES ARE DEFECTIVE ?

****

THE PROCESS CAN BE MODELED BY A BINOMIAL DISTRIBUTION. THE PROBABILITY THAT LESS THAN TWO FUSES ARE DEFECTIVE IS :

$$P(K < 2) = B(6,0.08,0) + B(6,0.08,1)$$

$$B(6,.08,0) = \binom{6}{0}(.08)^0(.92)^6 = 0.606$$

$$B(6,.08,1) = \binom{6}{1}(.08)^1(.92)^5 = 0.316$$

$$P(K < 2) = 0.922$$

5-15

A coin is thrown a number of times. Find the following probabilities: a) of obtaining exactly 2 heads & 2 tails in 4 throws, b) of obtaining at least one uninterrupted run of 3 heads in 5 throws, c) of predicting the outcome correctly in each of 5 throws.

****

(a) $\binom{4}{2}\left(\frac{1}{2}\right)^2\left(\frac{1}{2}\right)^2 = \frac{3}{8}$

(b) there are $2^5 = 32$ possible outcomes.

TO COUNT SUCCESSES: HHH (H OR T) (H OR T) ⇒ 4

T HHH (H OR T) ⇒ 2

(H OR T) T HHH ⇒ 2

$$\therefore P(\text{SUCCESS}) = \frac{4+2+2}{32} = \frac{1}{4}$$

(c) $\left(\frac{1}{2}\right)^5 = \frac{1}{32}$

5-16

Suppose that 5% of all items coming off a production line are defective. If 10 such items are chosen and inspected, what is the probability that at most 2 defectives are found?

****

G = GOOD $\quad Pr\{G\} = 0.95$

D = DEFECTIVE $\quad Pr\{D\} = 0.05$

$\quad\quad\quad\quad 1.00$

$n = 10$

$p = 0.05$ THIS IS A BINOMIAL PROBLEM.

$Pr\{\text{AT MOST } 2\ D\text{'s}\} = Pr\{0\ D\text{'s}\} + Pr\{1\ D\} + Pr\{2\ D\text{'s}\}$

$$= \binom{10}{0}(.05)^0(.95)^{10} + \binom{10}{1}(.05)^1(.95)^9 + \binom{10}{2}(.05)^2(.95)^8$$

$$= 0.5987 + 0.3151 + 0.0746$$

$$= 0.9885$$

5-17

If one tennis player has a 60% chance of winning any set he plays against a second player, what is the probability that the poorer player will win 3 sets of a 5 set match, if all 5 sets are played?

****

$P[3 \text{ WINS IN } 5 \text{ SETS}] = \binom{5}{3}(0.4)^3(0.6)^2$

$$= (10)(0.0640)(0.36) = 0.2304$$

FROM A STANDARD BINOMIAL TABLE,

$$b(3;5,0.4) = B(3;5,0.4) - B(2;5,0.4)$$

$$= 0.2304$$

5-18

GIVEN A "FAIR" COIN, WHAT IS THE PROBABILITY OF FLIPPING IT 10 TIMES AND OBTAINING 5 HEADS ?

****

THE INTUITIVE ANSWER TO MANY MAY BE 0.5. THAT IS WHAT MAKES PROBABILITY AND STATISTICS A DIFFICULT SUBJECT. OUR INTUITION IS OFTEN FAULTY IN PROBLEMS SUCH AS THIS. IT IS ALSO WHY GAMBLING ESTABLISHMENT MAKE MONEY. THE ODDS ARE IN THE HOUSE'S FAVOR.

THIS IS A SIMPLE BINOMINAL DISTRIBUTION PROBLEM :

THERE ARE 10 TRIALS
THE PROBABILITY OF SUCCESS IS 0.5
WE ARE LOOKING FOR <u>*EXACTLY*</u> 5 SUCCESSES !

$$B(N, P, K) = \binom{N}{K} (P)^K (1-P)^{N-K}$$

$$B(10, .5, 5) = \binom{10}{5} (.5)^5 (1-.5)^{10-5}$$

$$= \frac{10!}{5!\,5!} (.5)^5 (.5)^{.5}$$

$$= 0.246$$

IF THIS WAS AN "EVEN" BET, THE HOUSE WOULD WIN 3 TIMES OUT OF 4

5-19

The grades on a Statistics exam are Uniformly distributed between 50 and 100. Find the expected number of students who will receive a passing grade of 65 or better. There are 35 students in the class.

****

Let G = Grade on exam

$p = \Pr(\text{Passing grade})$
$= \Pr(G \geq 65)$
$= 35/50$
$= 0.7$

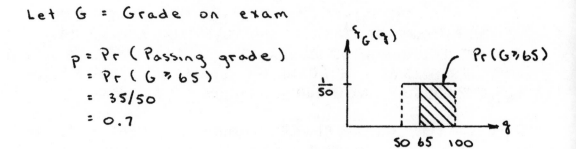

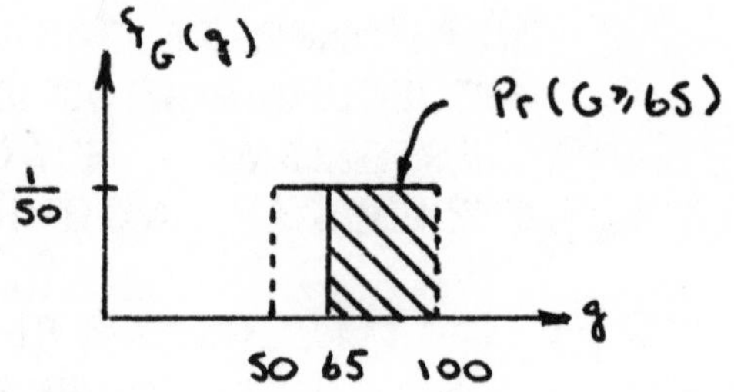

Let X = Number of students passing out of 35

Assuming student grades are independent, then X is binomially distributed with parameters $p = 0.7$ & $n = 35$

Expected number of passing grades $= \mu_X = np$
$= 35(0.7)$
$= 24.5$

Ans. 24 or 25 students

5-20

The probability that a certain rare blood disease will be fatal is 0.80. Find the probability that among a group of 15 patients having the disease.

(a) at least 9 will die.

(b) from 4 to 8 will die.

****

SITUATION IS BINOMIAL PROVIDED OUTCOMES ARE INDEPENDENT (PATIENTS TREATED AT DIFFERENT HOSPITALS BY DIFFERENT DOCTORS)

a) $Pr\,(\text{at least } 9 \mid n = 15,\ p_{death} = 0.80)$

$= Pr\,(\text{at most } 6 \mid n = 15,\ p_{surv.} = 0.20)$

$= B\,(6;\ 15,\ 0.20)$

$= 0.9819$

b) $Pr\,(4 \le X \le 8)$ where X = number of dead

$= Pr\,(7 \le Y \le 11)$ where Y = survivors

$= B\,(11;\ 15,\ 0.20) - B\,(6;\ 15,\ 0.20)$

$= 1.000 - 0.9819 = 0.0181$

MULTINOMIAL

5-21

Suppose 10 people each throw 2 coins. What is the probability that: a) three people throw two heads, three throw two tails, and four throw one head & one tail, b) no one threw a head and a tail.

****

a) BY MULTINOMIAL DISTRIBUTION:

$$\frac{10!}{3!\,3!\,4!}\left(\frac{1}{4}\right)^3\left(\frac{1}{4}\right)^3\left(\frac{1}{2}\right)^4 = \frac{175}{8192} = .0214$$

Where in tossing 2 coins: $P(2H) = P(2\text{ TAILS}) = \frac{1}{4}$

$P(1H \,\&\, 1T) = \frac{1}{2}$

(b) $$\left(\frac{1}{2}\right)^{10} = \frac{1}{1024}$$

5-22

A navigation system has four equally probable failure modes. Find the probability that among four different navigation systems,

a) each will fail in a different mode.

b) all will fail in the same mode.

****

Let X_i = Number of Type i failures among the 4 different systems $\quad i = 1, 2, 3, 4$

Then $\Pr(X_1 = x_1, X_2 = x_2, X_3 = x_3, X_4 = x_4)$, $\quad \sum_{i=1}^{4} x_i = n = 4$

$$= f_{X_1, X_2, X_3, X_4}(x_1, x_2, x_3, x_4)$$

$$= \frac{n!}{x_1!\, x_2!\, x_3!\, x_4!}\, p_1^{x_1} p_2^{x_2} p_3^{x_3} p_4^{x_4}, \quad \sum p_i = 1$$

(Multinomial Distribution)

a) $\Pr(X_1 = 1, X_2 = 1, X_3 = 1, X_4 = 1) = f_{X_1, X_2, X_3, X_4}(1, 1, 1, 1)$

$p_i = 0.25, \quad i = 1, 2, 3, 4$

$$= \frac{4!}{1!\,1!\,1!\,1!}(0.25)^4$$

$$= 0.09375$$

b) $\Pr(X_1 = 4, X_2 = X_3 = X_4 = 0) + \text{-------} + \Pr(X_1 = X_2 = X_3 = 0, X_4 = 4)$

$$= f_{X_1, X_2, X_3, X_4}(4, 0, 0, 0) + \text{-------} + f_{X_1, X_2, X_3, X_4}(0, 0, 0, 4)$$

$$= 4 f_{X_1, X_2, X_3, X_4}(4, 0, 0, 0)$$

$$= 4\left[\frac{4!}{4!\,0!\,0!\,0!}(0.25)^4 (0.25)^3\right]$$

$$= 0.01563$$

5-23

A very inexpensive pollster only takes samples of size 10. Suppose he asks 10 people a question for which the answer is yes, no, and undecided. What is the probability that he will get representative results if the true population percentages are 30, 50, and 20, respectively?

****

THE MULTINOMIAL DISTRIBUTION APPLIES.

$$f(3,5,2) = \frac{10!}{3!\,5!\,2!}(0.3)^3(0.5)^5(0.2)^2$$

$$= 0.085$$

HYPERGEOMETRIC

5-24

Widgets are produced in lots of forty. If a lot contains six defective widgets, what is the chance that a sample of five will contain two or more defective widgets? What is the expected number of defective widgets in a sample?

****

$$E(\text{DEFECTIVE WIDGETS}) = \frac{6}{40}(5) = 0.75 \quad \text{HYPERGEOMETRIC}$$

$$P(X \geq 2) = 1 - P(X < 2) = 1 - [P(X=0) + P(X=1)]$$

$$= 1 - \frac{\binom{6}{0}\binom{40-6}{5-0}}{\binom{40}{5}} - \frac{\binom{6}{1}\binom{40-6}{5-1}}{\binom{40}{5}}$$

$$= 1 - \frac{1\,(278256)}{658008} - \frac{6\,(46376)}{658008} = 15.42\ \%$$

5-25

Among ten applicants for four openings there are three females. The personnel director selected all three ladies plus one of the men. The other men filed a sex discrimination complaint. The director claimed that the selection was completely at random. What do you think?

****

SITUATION IS HYPERGEOMETRIC (SAMPLING WITHOUT REPLACEMENT) WHERE

$N = 10$ POPULATION SIZE

$n = 4$ SAMPLE SIZE

$a = 3$ SUCCESSES (FEMALES) IN POPULATION

$x = 3$ SUCCESSES FOUND IN SAMPLE

PROBABILITY OF RANDOM OCCURRENCE FOR THIS IS

$$h(x; n, a, N) = \frac{\binom{a}{x}\binom{N-a}{n-x}}{\binom{N}{n}}$$

$$= \frac{\binom{3}{3}\binom{7}{1}}{\binom{10}{4}} = \frac{\frac{3!}{3!\,0!}\quad\frac{7!}{1!\,6!}}{\frac{10!}{4!\,6!}}$$

$$= \frac{1}{30} = 0.033$$

PROBABILITY OF RANDOM OCCURRENCE IS SO LOW (JUST ABOUT 3%) THAT IT IS HARD TO BELIEVE. PERSONNEL DIRECTOR IS PROBABLY BIASED.

5-26

a) A lot of 10 articles contains 2 defectives. A random sample of 3 articles is to be selected. What are the resp. probabilities this random sample will contain 0, 1 and 2 defectives?
b) 1/5 of the articles in a continuous stream are defective. A random sample of 3 articles are to be selected. What are the resp. probs. of 0, 1 and 2 defectives in the group?

****

(a) $P(0) = \binom{8}{3} \Big/ \binom{10}{3} = \frac{7}{15}$

$P(1) = \frac{\binom{2}{1}\binom{8}{2}}{\binom{10}{3}} = \frac{7}{15}$

$P(2) = \frac{\binom{2}{2}\binom{8}{1}}{\binom{10}{3}} = \frac{1}{15}$

USING HYPERGEOMETRICAL PROBABILITIES

(b) $P(0) = \left(\frac{4}{5}\right)^3 = \frac{64}{125}$

$P(1) = 3 \cdot \left(\frac{4}{5}\right)^2 \cdot \frac{1}{5} = \frac{48}{125}$

$P(2) = 3 \cdot \left(\frac{4}{5}\right)\left(\frac{1}{5}\right)^2 = \frac{12}{125}$

USING BINOMIAL PROBABILITIES

5-27

A box contains 27 black and 3 red beads. A random sample of 5 beads is drawn without replacement. What is the probability the sample contained one red bead?

****

$$P(\text{one red bead}) = \frac{\binom{3}{1}\binom{27}{4}}{\binom{30}{5}} = \frac{\frac{3!}{2!\,1!} \cdot \frac{27!}{4!\,23!}}{\frac{30!}{5!\,25!}}$$

$$= \frac{75}{203} = .369$$

5-28

(a) A shipment of 15 concrete cylinders has been received by a contractor, five for a small project and the other ten for a larger project. Suppose that six of the 15 have a crushing strength which is below the specified minimum. If the five for the smaller project are randomly selected from the 15, compute the probability that four among the selected five have below the minimum crushing strength?

(b) What is the probability that at least one of the selected five has a below the minimum crushing strength?

(c) Suppose that out of the six that have a below the minimum crushing strength, four have a crushing strength which is acceptable to the contractor for yet another of his projects. The remaining two are too low to be of any use. Still selecting the five from the 15 cylinders, compute the probability that exactly 2 of the five are below strength but acceptable <u>and</u> one (of the five) is not acceptable.

****

(a) Let X denote the number of defective cylinders in the sample of five. X follows the hypergeometric distribution, with $N = 15$, $n = 5$ and $S = 6$ (S is the number of "successes" out of 15).

$$p(X=4) = \frac{\binom{6}{4}\binom{9}{1}}{\binom{15}{5}} = \frac{135}{3003} = \underline{\underline{0.045}}$$

(b) proba (at least one defective) $= p(X \geq 1) = 1 - p(X=0)$

$$= 1 - \frac{\binom{6}{0}\binom{9}{5}}{\binom{15}{5}} = \underline{\underline{0.958}}$$

(c) Out of the 6 defective cylinders, 4 are acceptable (for another project) and 2 are not acceptable

proba (2 acceptable defectives and 1 non-acceptable)

$$= \frac{\binom{4}{2}\binom{2}{1}\binom{9}{2}}{\binom{15}{5}} = \underline{\underline{0.144}}$$

■■ **5-29**

A bag of 25 jellybeans contains 4 black jellybeans. If 7 jellybeans are chosen at random in a single grasp from the bag, what is the probability that the handful will contain at least 2 black jellybeans?

****

Since sampling is without replacement the hypergenetic distribution is appropriate

$$P(x) = \frac{C_x^{n_1} \; C_{n-x}^{n_2}}{C_n^N} \qquad \text{where:} \quad \begin{array}{l} N = 25 \\ N_1 = 4 \\ N_2 = 17 \\ n = 7 \\ X = 2 \end{array}$$

$$P(2) = \frac{C_2^4 \; C_5^{17}}{C_7^{25}}$$

$$C_r^n = \frac{n!}{r!(n-r)!} \qquad C_2^4 = \frac{4!}{2!(2)!} = \frac{4 \cdot 3 \cdot \cancel{2}!}{2 \cdot \cancel{2}!} = 6$$

$$C_5^{17} = \frac{17!}{5!(12!)} = \frac{17 \cdot \cancel{16}^2 \cdot \cancel{15} \cdot 14 \cdot 13 \cdot \cancel{12}}{(\cancel{5} \cdot \cancel{4} \cdot \cancel{3} \cdot \cancel{2})(\cancel{12}!)} = 6188$$

$$C_7^{25} = \frac{25!}{7!(18!)} = \frac{25 \cdot \cancel{24} \cdot 23 \cdot 22 \cdot \cancel{21} \cdot \cancel{20}^2 \cdot 19 \cdot 18 \cdot 1}{(7 \cdot 6 \cdot 5 \cdot 4 \cdot 3 \cdot 2)(18!)} = 480700$$

$$P(2) = \frac{(6)(6188)}{480700} = .077 = 7.7\%$$

■■ **5-30**

Two identical canvas bags are filled with poker chips. The first bag has 70 green and 30 red chips; call this bag G. The second bag contains 30 green and 70 red chips; call this bag R. The chips are identical except for color. Choose one of the bags at random, and set the other aside. A random sample of 12 chips are chosen without replacement. Given that there are 8 green and 4 red chips in the sample, what is the probability that the bag is G?

****

$$Pr\{G \mid 4\text{ RED } \& \text{ } 8\text{ GREEN CHIPS}\} = \frac{Pr\{G\text{ BAG AND } 4\text{ RED } \& \text{ } 8\text{ GREEN}\}}{Pr\{4\text{ RED } \& \text{ } 8\text{ GREEN}\}}$$

$$= \frac{Pr\{4\text{ RED } \& \text{ } 8\text{ GREEN} \mid G\}\, Pr\{G\}}{Pr\{4\text{ RED } \& \text{ } 8\text{ GREEN} \mid G\}\, Pr\{G\} + Pr\{4\text{ RED } \& \text{ } 8\text{ GREEN} \mid R\}\, Pr\{R\}}$$

(BAYES' THEOREM)

$$= \frac{\left[\binom{30}{4}\binom{70}{8}\frac{1}{2}\right] / \binom{100}{12}}{\left[\binom{30}{4}\binom{70}{8}\frac{1}{2} + \binom{70}{4}\binom{30}{8}\frac{1}{2}\right] / \binom{100}{12}}$$

(HYPERGEOMETRIC C.D.F.)

$$= \left[1 + \frac{62!\,26!}{66!\,22!}\right]^{-1} = \left[1 + \frac{5 \times 23}{8 \times 21 \times 33}\right]^{-1}$$

$$= 0.9797$$

POISSON

5-31

Poisson probability distribution with mean μ is given by

$$p(n) = \frac{\mu^n e^{-\mu}}{n!} \qquad n = 0, 1, 2, \ldots$$

Show that

$$p(n+1) = \frac{\mu}{(n+1)} p(n)$$

A taxicab company in Cincinnati receives one call every 5 minutes on the average. If each trip normally requires 30 minutes of cab time, how many taxicabs does this company need in order to run a satisfactory operation? Assume Poisson distribution.

****

$$p(n+1) = \frac{\mu^{n+1} e^{-\mu}}{(n+1)!} = \frac{\mu}{(n+1)} \frac{\mu^n e^{-\mu}}{n!} = \frac{\mu}{(n+1)} p(n)$$

Let N = number of calls during a 5 min period. Assuming N to be Poisson distributed with mean $\mu = 1$, we have $p(n) = \frac{(1)^n e^{-1}}{n!} = \frac{1}{e\,n!}$

Hence

$$p(0) = 0.368$$
$$p(1) = \frac{p(0)}{1} = 0.368$$
$$p(2) = \frac{p(1)}{2} = 0.184$$
$$p(3) = \frac{p(2)}{3} = 0.061$$
$$p(4) = \frac{p(3)}{4} = 0.015$$
$$p(5) = \frac{p(4)}{5} = 0.003$$
$$p(6) = \frac{p(5)}{6} = 0.0005$$

Now a strategy has to be used to neglect small-probability events. Let us use the following:

Neglect $p(n+1)$ if it is less than or equal to 25% of $p(n)$.

Accordingly we should pick $n = 3$ as the operating number of cabs for a 5 min period. Since there are 6 five-minute periods in a half hour (the average trip duration), the company would need about

$6 \times 3 = \underline{\underline{18 \text{ taxicabs}}}$.

5-32

Accidents occur at an intersection at an average rate of 0.3 accidents per hour. Find the probability

a) of more than 4 accidents in 10 hours

b) that two consecutive accidents occur less than 3 hours apart

Note, the accidents occur as a Poisson process.

****

Let $X(t)$ = Number of accidents in t hours

$$\Pr[X(t) = K] = \frac{e^{-\alpha t}(\alpha t)^K}{K!} \qquad K = 0, 1, 2, \ldots$$

$$\alpha = 0.3 \frac{\text{accidents}}{\text{hour}}$$

a) $\Pr[X(10) > 4] = 1 - \Pr[X(10) \le 3]$

Note, $X(10)$ is a Poisson random variable with parameter $\lambda = \alpha t = 0.3(10) = 3$

$$= 1 - F(3;3)$$
$$= 1 - 0.6472$$
$$= 0.3528$$

b) Let T = Time between accidents

Then T is exponentially distributed, $f_T(t) = \alpha e^{-\alpha t}$

$$\Pr(T < 3) = \int_0^3 f_T(t)\,dt$$

$$= \int_0^3 \alpha e^{-\alpha t}\,dt$$

$$= -e^{-\alpha t}\Big|_0^3$$

$$= 1 - e^{-(0.3)(3)}$$

$$= 0.5934$$

5-33

The number of calls per minute arriving at a telephone switchboard may be assumed to be a random variable with a Poisson distribution. If an average of 300 calls per hour arrive at a switchboard, what is the probability that no more than 2 calls will arrive during any given minute?

****

λ = 300/HR = 5/MIN.

P[NO MORE THAN 2 CALLS]

= P[0] + P[1] + P[2]

FROM A STANDARD POISSON TABLE,

P[2 OR LESS] = 0.125

FROM THE PROBABILITY FUNCTION,

$$\frac{(5)^0 e^{-5}}{0!} + \frac{(5)^1 e^{-5}}{1!} + \frac{(5)^2 e^{-5}}{2!} = 0.1246$$

5-34

Assume that customers enter a store at the rate of 120 persons/hour. Assuming a Poisson distribution, a) What is the probability that during a 2 minute interval no one will enter the store? b) What time interval is such that the probability is ½ that no one will enter the store during that interval?

****

(a) 120/hr = 2/min = 4/2 min. $\therefore \lambda = 4$

FOR $\lambda = 4$, $P(x=0) = .0183$ (by Poisson tables)

(b) FOR X = 0, Probability is .50 for $\lambda = .69$

$\therefore \frac{.69}{2} = .345$ minutes = 20.7 seconds

5-35

The number of oil tankers, say N, arriving at a certain refinery each day has a Poisson distribution with parameter $\lambda = 2$. Present port facilities can service three tankers a day. If more than three tankers arrive in a day, the tankers in excess of three must be sent to another port.

a) On a given day what is the probability of having to send tankers away?
b) How much must present facilities be increased to permit handling of all tankers on about 90% of the days?
c) What is the expected number of tankers arriving per day?
d) What is the expected number of tankers serviced daily?
e) What is the expected number of tankers turned away daily?

****

a) $P\{\text{SENDING TANKERS AWAY}\} = P\{N \geq 4\}$; $N = \#$ TANKERS

$= 1 - \{P\{N=0\} + P\{N=1\} + P\{N=2\} + P\{N=3\}\}$

$= 1 - \{e^{-2} + 2e^{-2} + 2e^{-2} + 8e^{-2}/6\} = 1 - e^{-2}\{1 + 4 + 4/3\}$

$= 0.1429$ NOTE: $P\{N=k\} = e^{-2}2^k/k!$

b) $P\{\text{SEND TANKERS AWAY}\} \cong 0.1 \cong 1 - \sum_{i=0}^{k} P\{N=i\}$ FOR SOME k.

TRY $k=4$: $\sum_{i=0}^{4} P\{N=i\} = e^{-2}\{1 + 2 + 2 + 4/3 + 2/3\} = 7e^{-2}$

$1 - 7e^{-2} = 0.0527$. THUS, INCREASE FACILITIES ~ 25%.

c) $E[N] = \lambda = 2$

d) $N_S = \#$ SERVICED

$E[N_S] = \sum_{i=0}^{3} i P\{N_S = i\} = \sum_{i=0}^{2} i P\{N=i\} + 3P\{N \geq 3\}$

$= 0 \times e^{-2} + 1 \times 2e^{-2} + 2 \times 2e^{-2} + 3 \times (1 - 5e^{-2}) = 1.782$

e) $N_T = \#$ TURNED AWAY ; $N = N_S + N_T$

$E[N_T] = E[N] - E[N_S] = 2.0 - 1.782 = 0.218$

5-36

The number of breakdowns per day of a certain machine is thought to be Poisson distributed. During a sampling period of 100 days, 12 days had no breakdowns, 20 days had one breakdown, 33 days had two breakdowns, 16 days had three breakdowns, 10 days had four breakdowns, 6 days had five breakdowns, 2 days had six breakdowns, and 1 day had seven breakdowns. No day was observed when the machine broke down more than seven times. Determine if the data in fact follow a Poisson distribution at the 5% significance level.

****

The χ^2 (chi-square) test may be used to compare the observed frequencies to those expected from a Poisson distribution with

$$\text{Mean} = \mu = \frac{(12)(0)+(20)(1)+(33)(2)+(16)(3)+(10)(4)+(6)(5)+(2)(6)+(1)(7)}{100} = 2.23$$

If k represents the number of breakdowns in a given day, the Poisson probability of k breakdowns is given by

$$\frac{e^{-\mu}\mu^k}{k!}$$

. This is multiplied by 100 days for each $k = 0, 1, 2, \ldots$ to give the expected frequencies.

No. of breakdowns per day (k)	0	1	2	3	4	5	6	7	>7
Poisson probability of k breakdowns	0.1075	0.2398	0.2674	0.1987	0.1108	0.0494	0.0184	0.0059	0.0021
No. of observed days with k breakdowns	12	20	33	16	10	6	2	1	0
						9 (combined)			
No. of expected days with k breakdowns	10.75	23.98	26.74	19.87	11.08	4.94	1.84	0.59	0.21
						7.58 (combined)			

(Combine to avoid any frequency < 5) →

Degrees of freedom = 6 frequencies − 1 − 1 population parameter (μ)
= 4 df

$$\chi^2 = \frac{(12-10.75)^2}{10.75} + \frac{(20-23.98)^2}{23.98} + \frac{(33-26.74)^2}{26.74} + \frac{(16-19.87)^2}{19.87} + \frac{(10-11.08)^2}{11.08} + \frac{(9-7.58)^2}{7.58}$$

$$= 3.396 < 9.488 \text{ (table } \chi^2 \text{ for 5\% significance and 4 df)}$$

Conclusion: Data do not differ significantly from Poisson distribution.

5-37

The probability of a no-hitter being pitched in a major league ball game is 0.0005. Find the probability of at least 2 no-hitters during the season if there are 4 leagues of 6 teams and each team plays a 162 game schedule.

****

Number of games played $= \dfrac{4 \times 6 \times 162}{2} = 1944$

Let r.v. X = Number of no-hitters in a season

Then X is binomially distributed, $b(n,p)$
where $p = Pr(\text{no-hitter}) = 0.0005$
$n = 1944$ games per season

$$Pr(X \geq 2) = 1 - Pr(X < 2) = 1 - Pr(X \leq 1) = 1 - B(1; 1944, 0.0005)$$

where $B(1; 1944, 0.0005) = \sum_{i=0}^{1} b(i; 1944, 0.0005)$

Due to the large value of n and small p, the Poisson approximation to the binomial is justified

$$\lambda = np = (1944)(0.0005) = 0.972$$

$$B(1; 1944, 0.0005) \approx F(1; 0.972) \quad \text{Poisson c.d.f.}$$

Interpolating the Poisson c.d.f. table values

0.95	0.754
0.972	0.754 + x
1.00	0.736

$$\frac{0.972 - 0.95}{1.00 - 0.95} = \frac{x}{0.736 - 0.754}$$

$x = -0.00792$

$F(1; 0.972) = 0.746$

$$Pr(X \geq 2) \approx 1 - 0.746 = 0.254$$

The exact answer is obtained by evaluating the binomial probability distribution function.

$$Pr(X \le 1) = Pr(X=0) + Pr(X=1)$$
$$= b(0; 1944, 0.0005) + b(1; 1944, 0.0005)$$

where $b(x; n, p) = \frac{n!}{x!\,(n-x)!} p^x (1-p)^{n-x}$, $x = 0, 1, 2, \ldots$

$$Pr(X \le 1) = \frac{(1944)!}{0!\,(1944)!}(0.0005)^0(0.9995)^{1944}$$

$$+ \frac{(1944)!}{1!\,(1943)!}(0.0005)^1(0.9995)^{1943}$$

$$= 0.746$$

$$Pr(X \ge 2) = 1 - 0.746 = 0.254$$

5-38

Between 8:30 AM and 9:30 AM cars arrive in the college parking lot in a Poisson stream with a mean of 10 cars per minute. What is the probability that during any randomly chosen minute between 8:30 AM and 9:30 AM exactly 11 cars will arrive in the lot?

****

$P(x)\ \frac{\lambda^x e}{x!}$ $\quad \lambda = 10 \quad X = 11$

$$P(11) = \frac{(10)^{11} e^{-10}}{11!} = \frac{10^{11} \times 4.54 \times 10^{-5}}{3.99 \times 10^7} = \frac{4.54}{3.99} \times 10^{-1} = .1138$$

$P(11) = .0125$?

5-39

The number of passengers arriving at a particular bus stop is Poisson distributed with a mean rate of 1.2 passengers per minute.

a) Calculate the probability that more than 4 passengers arrive in the next 3 minutes.

b) Calculate the expected number of passenger arrivals in a 5 minute interval. Also calculate the standard deviation of the number of arrivals in 5 minutes.

c) If we are told that 15 passengers have arrived between 7:00 A.M. and 7:10 A.M., calculate the conditional probability that 10 out of the 15 passengers arrived in the last 5 minutes (i.e., between 7:05 and 7:10 A.M.).

****

Let N_T denote the number of passengers arriving in T minutes.
N_T is Poisson distributed with parameter $\lambda_T = 1.2\,T$ passengers.

a) $p(N_T > 4) = 1 - p(N_T \leq 4)$

$$= 1 - \left[p(N_T=0) + p(N_T=1) + p(N_T=2) + p(N_T=3) + p(N_T=4) \right]$$

$$= 1 - e^{-3.6} - 3.6e^{-3.6} - \frac{(3.6)^2 e^{-3.6}}{2} - \frac{(3.6)^3 e^{-3.6}}{2 \times 3} - \frac{(3.6)^4 e^{-3.6}}{2 \times 3 \times 4}$$

$$= 1 - 0.706 = \underline{\underline{0.294}}.$$

b) $E[N_5] = \lambda_5 = 5 \times 1.2 = 6$ passengers.

$V[N_5] = \lambda_5 = 6 \Rightarrow$ std. dev. $= \sqrt{6} = 2.45$ passengers.

c) Define the following events: $A = \{15$ pax. between 7:00 – 7:10 A.M.$\}$
$B = \{10$ pax. between 7:05 – 7:10 " $\}$
$C = \{5$ pax. " 7:00 – 7:05 " $\}$

need $p(B|A) = \dfrac{p(B \cap A)}{p(A)}$

Noting that $\{B \cap A\} = \{B \cap C\}$, and that B and C are <u>independent</u>,

$$p(B|A) = \frac{p(B \cap C)}{p(A)} = \frac{p(B) \cdot p(C)}{p(A)} = \frac{p(N_5=10) \cdot p(N_5=5)}{p(N_{10}=15)}$$

$p(N_5 = 5) = \dfrac{(6)^5 e^{-6}}{5!} = 0.1606$; $p(N_5 = 10) = \dfrac{(6)^{10} e^{-6}}{10!} = 0.0413$;

$p(N_{10} = 15) = \dfrac{(12)^{15} e^{-12}}{15!} = 0.0724$ $\quad\therefore p(B|A) = \underline{\underline{0.092}}$

NEGATIVE BINOMIAL

5-40

Based on past experience, a baseball player has a 1 in 4 chance of getting a hit every time up. Find the probability of him making less than 4 outs before he gets his 2 nd hit.

****

Let r.v. X = Number of outs made before 2 nd hit
$p = Pr$ (Hit each time up)

Then X follows a negative binomial distribution,

i.e. $f_X(x) = \binom{x+r-1}{r-1} p^r (1-p)^x \qquad x = 0, 1, 2, ----$

where $r = 2$ and $p = 0.25$

$$Pr(X < 4) = Pr(X \leq 3)$$

$$= \sum_{x=0,1,2,3} \binom{x+2-1}{2-1} (0.25)^2 (0.75)^x$$

$$= (0.25)^2 \sum_{x=0,1,2,3} \binom{x+1}{1} (0.75)^x$$

$$= (0.25)^2 \sum_{x=0,1,2,3} (x+1)(0.75)^x$$

$$= (0.25)^2 \left[1 + 2(0.75) + 3(0.75)^2 + 4(0.75)^3 \right]$$

$$= 0.367$$

GEOMETRIC

5-41

An evenly balanced coin is tossed. Find the probability that the number of tosses made in arriving at the first tail is:

(a) Exactly 2

(b) At least 1

(c) From 1 to 3 inclusively

(d) Greater than 2

****

(a) $$P(K) = P(1-P)^{K-1}, \quad P = 0.5 = P(H) = P(T)$$

$k = 1, 2, \ldots.$

$$P(K=2) = 0.5(1-0.5)^{2-1} = 0.25$$

(b) $$P\{K \geq 1\} = \sum_{k=1}^{\infty} P(K) = 1$$

(c) $$P\{1 \leq k \leq 3\} = \sum_{k=1}^{3} 0.5(0.5)^{k-1}$$

$$= 0.5(1) + 0.5(0.5) + 0.5(0.5)^2$$

$$= 0.875$$

(d) $$P\{K > 2\} = \sum_{k=3}^{\infty} P(k) = 1 - \sum_{k=1}^{2} P(k)$$

$$P\{k > 2\} = 1 - 0.5 - 0.25 = 0.25$$

5-42

a) It has been determined that the probability that the travel time experienced by a driver going from Austin to Houston along U.S. Highway 290 exceeds 4 hours is equal to 0.13. The travel times of 7 different drivers going from Austin to Houston along U.S. Highway 290 are measured. Assuming that the respective travel times are independent, calculate the probability that the travel times of at least 3 of the 7 drivers will be in excess of 4 hours.

b) Some drivers, after arriving in Houston may decide to continue to Galveston. Suppose that we were to ask every driver arriving in Houston from Austin (via U.S. Hwy 290) if he/she will be continuing on to Galveston. Assume further that 5% of all such drivers continue on to Galveston.

What distribution is appropriate to calculate the probability that at most 30 drivers have to be stopped before finding one that is continuing to Galveston? Justify your answer by clearly defining the random variable in question and stating the assumptions underlying its distribution.

****

a) Let N denote the number of drivers out of 7 with travel time in excess of 4 hours.

N binomially distributed, with $n = 7$ and $p = 0.13$.

$$p(N \geq 3) = 1 - p(N \leq 2) = 1 - [p(N=2) + p(N=1) + p(N=0)]$$

$$= 1 - \binom{7}{2}(0.13)^2(0.87)^5 - \binom{7}{1}(0.13)(0.87)^6 - \binom{7}{0}(0.87)^7$$

$$= 1 - 0.177 - 0.395 - 0.377 = \underline{0.051}.$$

b) Let W denote the number of drivers that have to be stopped before finding one continuing to Galveston.

W is <u>geometrically distributed</u>, since each driver can be viewed as an independent Bernoulli trial ("success" = "continuing") with constant probability of success across trials ($p = 0.05$). Thus W can be viewed as the number of trials until first success.

$$p(W \leq 30) = \sum_{i=1}^{30} (1-p)^{i-1} p = \sum_{i=1}^{30} (0.95)^{i-1} (0.05).$$

5-43

Three people toss a coin and the odd man pays for the coffee. If the coins all turn up the same, they are tossed again. Find the probability that fewer than four tosses are needed.

****

Let X = Number of tosses until not all the same

q = Pr (All the same on any toss)
= Pr (HHH OR TTT)
= Pr (HHH) + Pr (TTT)
= 1/8 + 1/8
= 1/4

Note, Pr (HHH) = Pr(H) Pr(H) Pr(H) assuming independent tosses

Therefore, p = Pr (Not all the same on any toss)
= 3/4

$$Pr(X = x) = f_X(x) = p\,q^{x-1} \qquad x = 1, 2, 3, \ldots$$

$$\begin{aligned} Pr(X < 4) &= Pr(X \le 3) \\ &= f_X(1) + f_X(2) + f_X(3) \\ &= pq^0 + pq^1 + pq^2 \\ &= p(1 + q + q^2) \\ &= \frac{3}{4}\left[1 + \frac{1}{4} + \frac{1}{16}\right] \\ &= \frac{63}{64} \end{aligned}$$

5-44

Cars arriving at a particular highway-patrol safety checkpoint form a Poisson process with mean rate of 100 vehicles per hour. Each car is tested for mechanical defects, and the drivers are informed of needed repairs. Suppose that 10% of all cars in the state need such repairs.

a) What distribution is appropriate for finding the probability of exactly 5 cars in need of repair out of the next 10 cars that arrive at the checkpoint? Compute that probability.

b) Compute the probability that at most two vehicles are in need of repairs out of the next 10 arrivals.

c) What distribution is appropriate for finding the probability that the patrol will inspect 50 cars until they find one in need of repair? Compute that probability.

****

a) Each car can be viewed as an independent Bernoulli trial ("success" ⇔ "in need of repair"), with constant probability of success across trials ($p = 0.1$).
Let X denote the number of cars in need of repairs out of the next 10 cars ($X \Leftrightarrow$ # of successes in 10 trials).
X is thus binomially distributed (with $n = 10$, $p = 0.1$)

$$p(X=5) = \binom{10}{5}(0.1)^5(0.9)^5 = \frac{10!}{5!\,5!}(0.1)^5(0.9)^5 = \underline{\underline{0.0015}}$$

b)
$$p(X \leq 2) = p(X=0) + p(X=1) + p(X=2)$$

$$= \binom{10}{0}(0.1)^0(0.9)^{10} + \binom{10}{1}(0.1)^1(0.9)^9 + \binom{10}{2}(0.1)^2(0.9)^8$$

$$= \underline{\underline{0.93}}$$

c) Let Y denote the number of cars inspected until the first "success".
Y is geometrically distributed;

$$p(Y=50) = (0.9)^{49}(0.1) = \underline{\underline{5.7 \times 10^{-4}}}$$

5-45

Denote the probability of success of a trial by p and the probability of failure by q. If the trial is repeated successively, the random variable N representing the number of trials to the first success obeys the geometric probability distribution

$$p(n) = q^{n-1} p$$

Show that

$$\sum_{n=1}^{\infty} p(n) = 1$$

What is the average number of trials to the first success?

A CNC (Computer Numerical Control) machine is programmed to cut gear wheels. The probability of not satisfying the manufacturing tolerance in each finished item is 0.1. An automatic gaging device checks the accuracy of the gear wheels and stops the machine if it is not satisfactory. What is the probability that the milling machine has to be stopped for readjustment after cutting 5 gear wheels?

****

$$\sum_{n=1}^{\infty} p(n) = \sum_{n=1}^{\infty} q^{n-1} p = p \sum_{n=1}^{\infty} q^{n-1}$$

This is a geometric series with the common ratio q satisfying $0 < q < 1$. It follows that

$$\sum_{n=1}^{\infty} q^{n-1} = \frac{1}{1-q} = \frac{1}{p} \qquad \text{because } p+q=1$$

Hence $\displaystyle\sum_{n=1}^{\infty} p(n) = p \cdot \frac{1}{p} = \underline{\underline{1}}$

Mean (expected value) of the random variable N is

$$E(N) = \sum_{n=1}^{\infty} n\, p(n) = \sum_{n=1}^{\infty} n q^{n-1} p = p \sum_{n=1}^{\infty} n q^{n-1}$$

Now since $\displaystyle\sum_{n=0}^{\infty} q^n = \frac{1}{1-q}$ {The sum of a geometric series

by differentiating this relation with respect to q;

$$\sum_{n=1}^{\infty} n q^{n-1} = \frac{1}{(1-q)^2} = \frac{1}{p^2} \qquad \{\text{because } 1-q = p$$

Hence $E(N) = \frac{p}{p^2} = \underline{\underline{\frac{1}{p}}}$

This is the average number of trials to the first success.

For the given problem, $p = 0.1$ and $n = 5$.

Hence

$$p(5) = (1-0.1)^{5-1} \times 0.1 = \underline{\underline{0.06561}}$$

This probability is very small for most practical purposes.

5-46

A simple gambling game consists of flipping a coin until a head results. If you pay $.50 per flip and get $1.00 when a head results, what is the probability that you will lose $3.00 on any single game?

****

THE GEOMETRIC DISTRIBUTION APPLIES.

LOSING $3.00 MEANS THAT THE FIRST HEAD APPEARS ON THE EIGHTH TOSS.

$$P[\text{1ST HEAD ON 8TH TOSS}] = p(1-p)^7$$

$$= (0.5)(0.5)^7 = 0.00391$$

OTHER

5-47

Let x be a random variable with the following probability distribution:

x	-3	6	9
P(X=x)	1/6	1/2	1/3

Find E(x) and $E(x^2)$ and then, using the laws of expectation, evaluate:

(a) $E((2x+1)^2)$

(b) $E(x-E(x))^2$.

****

$$E(x) = -3(1/6) + 6(1/2) + 9(1/3) = 33/6 = 11/2$$

$$E(x^2) = 9(1/6) + 36(1/2) + 81(1/3) = \frac{279}{6} = \frac{73}{2}$$

(a) $$E\{(2x+1)^2\} = E\{4x^2 + 4x + 1\}$$
$$= 4E(x^2) + 4E(x) + 1$$
$$= 4(73/2) + 4(11/2) + 1 = 169$$

(b) $$E(x-E(x))^2 = E\{x^2 - 2xE(x) + E^2(x)\}$$
$$= E(x^2) - 2E^2(x) + E^2(x)$$
$$= E(x^2) - E^2(x)$$
$$= 73/2 - (11/2)^2$$
$$= 73/2 - 121/4 = 25/4$$

5-48

The following is a list of processes whose outcomes can be described by a random variable. For each of them, select the distribution that will best represent the probability of occurrence of the corresponding random variable.

a) The number of cars arriving at a gas service station between 2 and 4 p.m.

____

b) The interarrival time between cars arriving at a gas service station between 2 and 4 p.m.

____

c) The number of heads that will occur in three tosses of a fair coin.

____

d) The selection of any given number on a lottery's "daily number"

____

e) The occurrence of the first "seven" in a game of craps (two fair die are tossed, the points in each face up are added)

____

f) The results of a nationally administered test to enter graduate school

____

g) One "volunteer" selected by writing the name on a slip of paper for each eligible person and then asking the first passerby to pull one of the slips from a hat

____

h) Given that a family has 8 children, the distribution of the numbers of boys

____

i) The second juror accepted was the seventh person interviewed

____

****

a. Poisson b. Exponential c. Binomial

d. Discrete uniform e. Geometric

f. Normal g. Uniform h. Binomial

i. Negative binomial

5-49

Consider the two discrete random variables X and Y. The marginal cumulative distribution function (cdf) of X is given by:

$$F_X(x) = \begin{cases} 0.0 , & x < 0 \\ 0.5 , & x = 0 \\ 0.8 , & x = 1 \\ 1.0 , & x = 2 \end{cases}$$

a) What is the probability mass function of X?

b) The conditional probability mass function (pmf) of Y given that X = 0 is given by:

$$f_{Y|X}(y|0) = \begin{cases} 0.2 , & y = 1 \\ 0.3 , & y = 2 \\ 0.1 , & y = 3 \\ 0.4 , & y = 4 \\ 0 & \text{elsewhere} \end{cases}$$

Compute the values of the joint pmf of X and Y for the missing (X,Y) pairs and complete the joint pmf table below:

X \ Y	1	2	3	4
0				
1	0.00	0.10	0.15	0.05
2	0.10	0.05	0.0	0.05

c) Using the above probability table, find the probability that the sum (X + Y) exceeds 5.

d) Compute E[Y], E[X] and V[X]

****

a) $$f_X(x) = \begin{cases} 0.5, & x = 0 \\ 0.3, & x = 1 \\ 0.2, & x = 2 \\ 0 & \text{elsewhere} \end{cases}$$

b) $f_{X,Y}(0, y) = f_{Y|X}(y|0)\ f_X(0) = f_{Y|X}(y|0) \times 0.5$,

resulting in the following values in the table:

X \ Y	1	2	3	4
0	0.10	0.15	0.05	0.20

c) $p([X+Y] > 5) = f_{X,Y}(2,4) = \underline{\underline{0.05}}$

d) $E[Y] = \sum_{y=1}^{4} y\, f_Y(y)$,

where $f_Y(y)$ is the marginal pmf of Y;

it can be obtained from the joint pmf $f_{X,Y}(x,y)$ as follows:

$$f_Y(y) = \sum_{x=0}^{2} f_{X,Y}(x,y),$$

yielding

$$f_Y(y) = \begin{cases} 0.20, & y = 1, 3 \\ 0.30, & y = 2, 4 \\ 0 & \text{elsewhere} \end{cases}$$

thus $E[Y] = 1 \times 0.2 + 2 \times 0.3 + 3 \times 0.2 + 4 \times 0.3 = \underline{\underline{2.6}}$

$$E[X] = \sum_{x=0}^{2} x\, f_X(x) = 1 \times 0.3 + 2 \times 0.2 = \underline{\underline{0.7}}$$

$$V[X] = E[X^2] - E^2[X]$$

$$= \sum_{x=0}^{2} x^2 f_X(x) - (0.7)^2$$

$$= 1 \times 0.3 + 4 \times 0.2 - 0.49 = \underline{\underline{0.61}}$$

5-50

Vehicles arrive randomly at an intersection at an average rate of 10 per minute. Traffic counts at this location indicate that 15 percent of these vehicles are trucks.

a) Assuming Poisson arrivals, calculate the probability that exactly 15 vehicles arrive in a 2-minute interval.

b) What is the mean and the variance of the number of trucks that arrive in a 3-hour period?

c) What distribution is appropriate for finding the probability that at least 6 vehicles arrive at the intersection before the arrival of a truck? Compute that probability.

d) What distribution is appropriate for finding the probability that there will be at least 3 trucks in the next 20 vehicle arrivals. Compute that probability.

e) Let X be the number of trucks in the next 50 arrivals; we define the random variable Y = 500 - 10X; what is the expected value of Y? What is its standard deviation?

f) Compute the probability that the next 25 vehicles are such that 15 arrive in the next 3 minutes, followed by an interruption of 30 seconds which is subsequently followed by the arrival of the remaining 10 vehicles in 2 minutes (with no other vehicles arriving in these 2 minutes).

****

Let N_T denote the number of vehicles that arrive in T minutes; N_T is Poisson distributed with parameter $\lambda_T = 10T$; thus $p(N_T = x) = \frac{(10T)^x e^{-10T}}{x!}$

a) $p(N_2 = 15) = \frac{(20)^{15} e^{-20}}{15!} = 0.052$

b) The rate of truck arrivals is $0.15 \times 10 = 1.5$ trucks/minute. If M_T denotes the number of trucks that arrive in T minutes,

then M_T will be Poisson distributed with parameter $1.5T$.

$$E[M_{180}] = 1.5 \times 180 = 270.0 \text{ trucks}$$

$$V[M_{180}] = 1.5 \times 180 = 270.0 \text{ (trucks)}^2$$

c) Let K denote the number of vehicles that arrive before the arrival of a truck + 1.

K can thus be viewed as the number of trials until the first success (in a sequence of independent Bernoulli trials).
K is thus <u>geometrically distributed</u>.

proba (at least 6 vehicles arrive <u>before</u> truck arrival) $= p(K \geq 7)$

$$= 1 - \sum_{i=1}^{6} p(K=i) = 1 - \sum_{i=1}^{6} (0.85)^{i-1}(0.15) = \underline{0.377}$$

d) Let W denote the number of trucks in the next 20 arrivals.

W is <u>binomially distributed</u>, with $n=20$ and $p=0.15$.

$$p(W \geq 3) = 1 - p(W=0) - p(W=1) - p(W=2)$$

$$= 1 - \binom{20}{0}(0.15)^0(0.85)^{20} - \binom{20}{1}(0.15)(0.85)^{19} - \binom{20}{2}(0.15)^2(0.85)^{18}$$

$$= \underline{\underline{0.595}}$$

e) $E[Y] = 500 - 10\,E[X] = 500 - 10\,(50 \times 0.15) = 425$

$V[Y] = 100\,V[X] = 100\,[50 \times 0.15 \times 0.85] = 637.5$

std. dev. $= \sqrt{637.5} = 25.25$.

f)

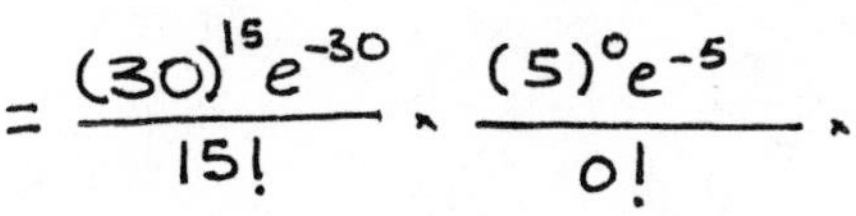

$$\text{proba} = p\left[(N_3 = 15) \cap (N_{0.5} = 0) \cap (N_2 = 10)\right]$$

$$= p(N_3 = 15) \times p(N_{0.5} = 0) \times p(N_2 = 10)$$

$$= \frac{(30)^{15} e^{-30}}{15!} \times \frac{(5)^0 e^{-5}}{0!} \times \frac{(20)^{10} e^{-20}}{10!}$$

$$= 0.001 \times 0.0067 \times 0.0058 = 4.0 \times 10^{-8} \approx 0.$$

6 CONTINUOUS DISTRIBUTIONS

UNIFORM

6-1

A useful random number in stochastic simulation is a uniform (0,1), that is, a value of a random variable having the uniform distribution over the unit interval.

a) What is the density function for a uniform (0,1)?

b) What is the probability that a uniform (0,1) takes on a value of less than 0.3 or greater than 0.9?

****

a) $$f(x) = \begin{cases} \frac{1}{\beta - \alpha} & \text{FOR } \alpha < x < \beta \\ 0 & \text{ELSEWHERE} \end{cases}$$

$$= \begin{cases} 1 & \text{FOR } 0 < x < 1 \\ 0 & \text{ELSEWHERE} \end{cases}$$

b) $$P[(x < 0.3) \text{ OR } (x > 0.9)]$$

$$= 0.3 + 0.1 = 0.4$$

6-2

Random variables X and Y are Uniformly distributed over the intervals (0,A) and (0,B) respectively. The variance of Y is four times the variance of X. Show that the mean of Y is equal to "A".

****

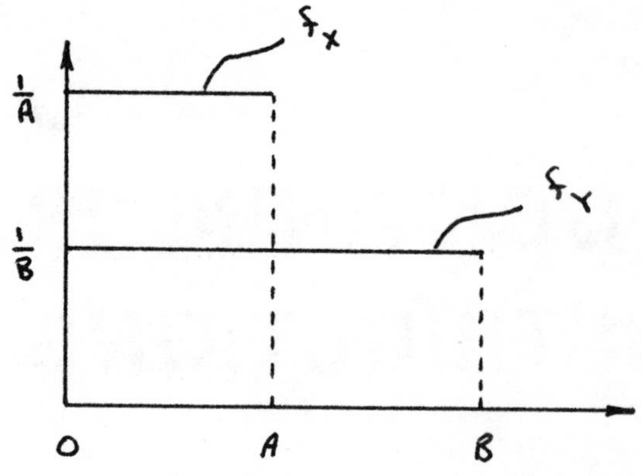

$$\sigma_Y^2 = 4\,\sigma_X^2$$

$$\frac{B^2}{12} = 4\,\frac{A^2}{12}$$

$$B = 2A$$

$$\mu_Y = \frac{B}{2} = A$$

TRIANGULAR

6-3

Random variable Y has probability density function

$$f_Y(y) = \begin{cases} Ay & 0 \le y \le B \\ 0 & \text{otherwise} \end{cases}$$

Find A and B if the mean of Y is 2.

****

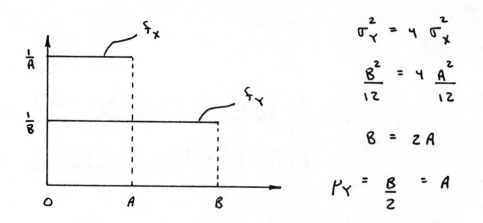

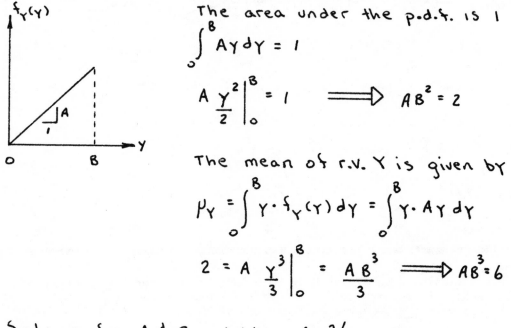

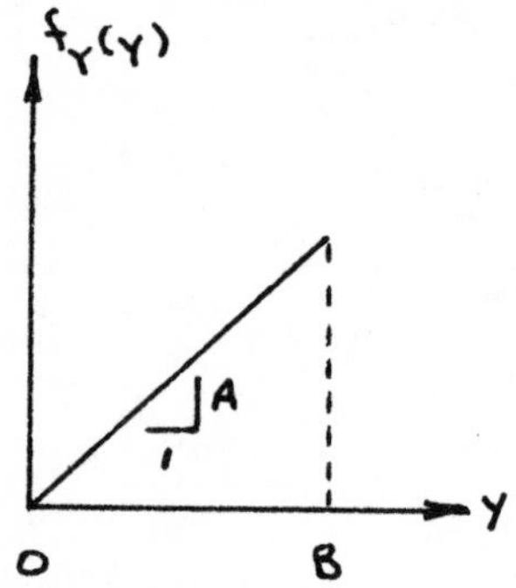

The area under the p.d.f. is 1

$$\int_0^B Ay\,dy = 1$$

$$A\,\frac{y^2}{2}\Big|_0^B = 1 \implies AB^2 = 2$$

The mean of r.v. Y is given by

$$\mu_Y = \int_0^B y \cdot f_Y(y)\,dy = \int_0^B y \cdot Ay\,dy$$

$$2 = A\,\frac{y^3}{3}\Big|_0^B = \frac{AB^3}{3} \implies AB^3 = 6$$

Solving for A & B yields $A = 2/9$, $B = 3$

NORMAL

6-4

If the diameters of 300 ball bearings were measured and were normally distributed with a mean of 0.452 cm and a standard deviation of 0.010 cm,

a. How many ball bearings would be expected to be smaller than 0.4425 cm? and

b. Seventy percent of the ball bearings would be expected to have a diameter greater than what value?

****

a.

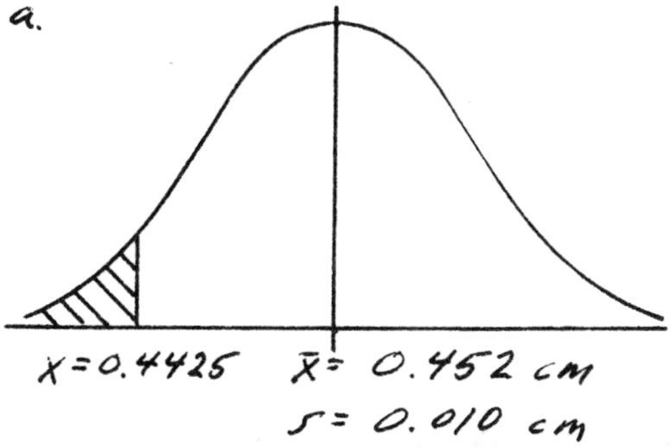

b.

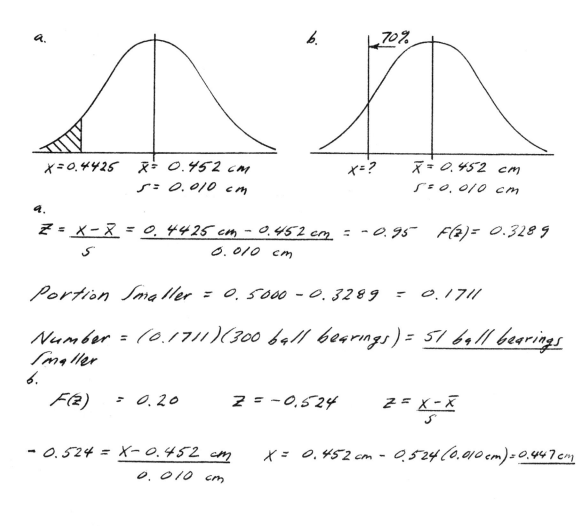

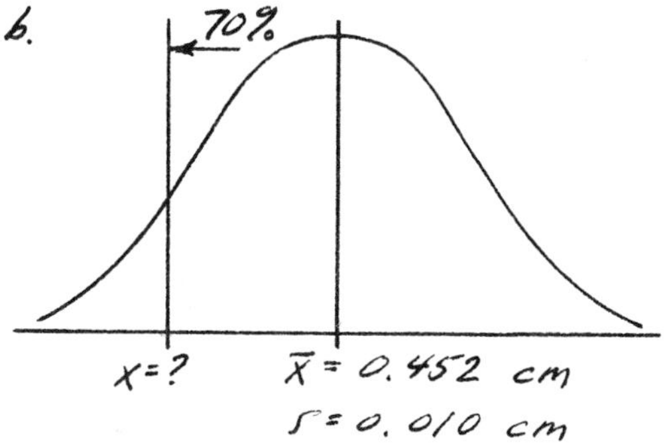

a.

$$Z = \frac{X - \bar{X}}{s} = \frac{0.4425 \text{ cm} - 0.452 \text{ cm}}{0.010 \text{ cm}} = -0.95 \qquad F(z) = 0.3289$$

Portion Smaller = 0.5000 − 0.3289 = 0.1711

Number = (0.1711)(300 ball bearings) = <u>51 ball bearings</u> Smaller

b.

$$F(z) = 0.20 \qquad Z = -0.524 \qquad Z = \frac{X - \bar{X}}{s}$$

$$-0.524 = \frac{X - 0.452 \text{ cm}}{0.010 \text{ cm}} \qquad X = 0.452 \text{ cm} - 0.524(0.010 \text{ cm}) = \underline{0.447 \text{ cm}}$$

6-5

The score a randomly selected person makes on an IQ test may be considered to be a value from a normal distribution with mean 100 and standard deviation of 15.

a) What is the probability that a randomly selected person will score less than 80 on an IQ test?

b) If a genius is defined to be a person who scores 145 or more on an IQ test, what percentage of the population are of genius level?

****

a) $P[IQ < 80] = P\left[Z < \frac{80-100}{15}\right]$

$= P[Z < -1.33]$

FROM A STANDARD NORMAL TABLE,

$= 1 - 0.9082 = 0.0918$

b) $P[IQ > 145] = P[Z > 3]$

$= 1 - 0.9987 = 0.0013$

$= .13\%$

6-6

A stamping machine produces can tops whose diameters are normally distributed with a standard deviation of 0.01 inches. At what nominal (or mean) diameter should the machine be set so that not more than 5% of the can tops produced have diameters exceeding 3 inches?

****

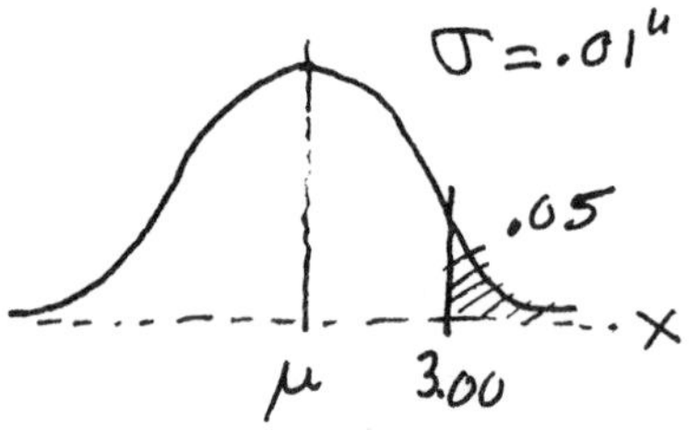

FOR AN AREA OF .05, $Z = 1.645$

$\therefore \frac{3-\mu}{\sigma} = 1.645$

AND $\mu = 3.00 - 1.645(.01)$

$= 3.00 - .0165$

$= 2.984''$

6-7

Repeated random samples of 100 bars of soap are weighed and the total weight Y found to be a random variable with μ_y = 50 lb. & $\sigma^2{}_y$ = 1 lb. If X, the weight of one bar of soap, is normally distributed, what is P(x < 0.4 lb).

******* **

$Y = X_1 + X_2 + \cdots\cdots + X_{100}$ Where $\mu_y = 50$, $\sigma^2_y = 1$

BUT $\mu_y = 100\,\mu_x$ ∴ $\mu_x = \frac{1}{2}$ lb.

AND $\sigma^2_y = 100\,\sigma^2_x$ ∴ $\sigma^2_x = .01$ and $\sigma_x = .1$ lb.

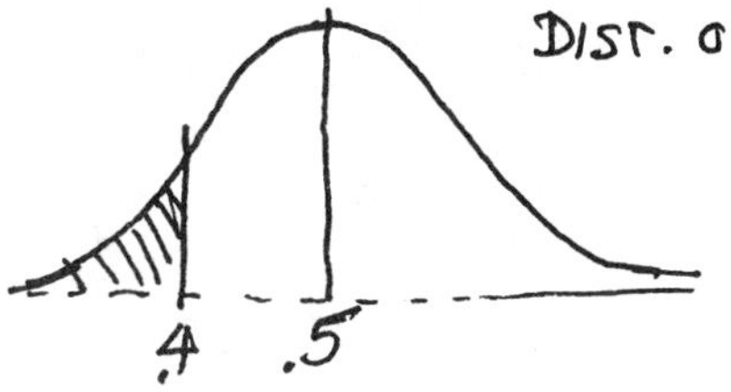

DIST. OF INDIV. WEIGHTS

$P(X < 0.4) = P\left(Z < \frac{.4 - .5}{.1}\right) = P(Z < -1)$

$= .1587$

6-8

a) The finished diameter on armored electric cable is normal with mean .775", standard deviation .010". Find the probability the diameter will exceed 0.790".

b) Assuming σ remains .010", at what value would the mean be set if the probability of exceeding .790" is to be 0.01?

****

(a)

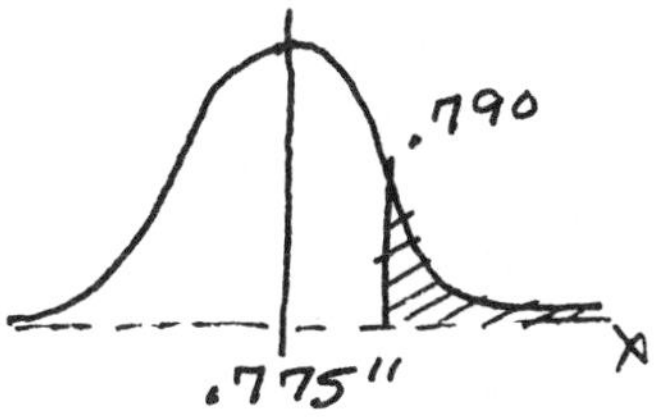

$Z = \frac{.790 - .775}{.010} = 1.5$

AREA ABOVE $Z = 1.5$ IS .0668

∴ $P(X > .790") = .0668$

(b)

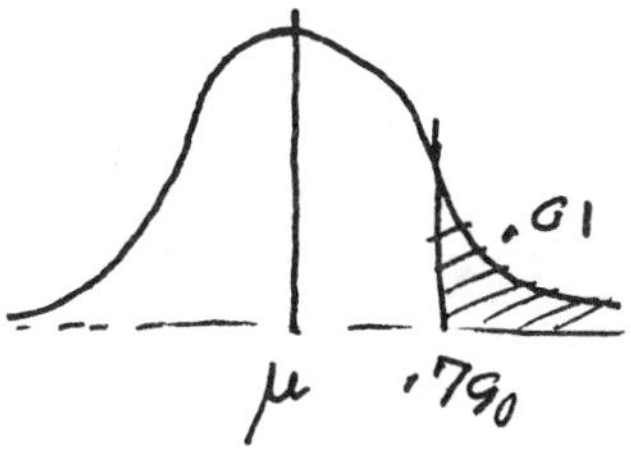

FOR AREA OF .01, $Z = 2.326$

∴ $\frac{.790" - X}{.010"} = 2.326$

then $X = \mu = .7667"$

6-9

A BATTERY IS USED IN A CRITICAL PIECE OF EQUIPMENT. IT IS REQUIRED TO DELIVER A MINIMUM OF 21 VOLTS. THE NOMINAL VOLTAGE IS 24 VOLTS. PAST TESTING HAS SHOWN THAT THE VARIANCE ON THE VOLTAGE IS 2.25.

IF A BATTERY IS SELECTED AT RANDOM, WHAT IS THE PROBABILITY THAT IT WILL BE UNSATISFACTORY ?

****

THE BATTERY IS UNSATISFACTORY IF ITS VOLTAGE IS LESS THAN 21 VOLTS. THE DISTRIBUTION OF VOLTAGES CAN BE MODELED BY THE NORMAL DISTRIBUTION.

$X = 21 \qquad \mu = 24 \qquad \sigma = 1.5$

REMEMBER, WE NEED TO WORK WITH STD. DEV HERE !

$Z = (21 - 24) / 1.5 = -2$

PROB (X <= 21) = 0.0228

6-10

The lifetime of a computer circuit board can be estimated by a normal c.d.f. with a mean of 2500 hours. If 95% of the boards are to last at least 2400 hours, what is the largest variance that the c.d.f. can have?

****

$T =$ LIFETIME ; $T \sim N(2500, \sigma^2)$

$Pr\{T > 2400\} = 0.95$ IMPLIES

$$Pr\left\{\frac{T-\mu}{\sigma} > \frac{2400-2500}{\sigma}\right\} = Pr\{Z > -100/\sigma\} = .95$$

THUS, $-\frac{100}{\sigma} = -1.645$; $\sigma = 60.8$; $\sigma^2 = 3695\ hr^2$

6-11

In the spring each year the applicants to a certain college take a qualifying examination to determine their eligibility for admission to a special program. Only the top 25% of those taking the exam are eligible for admission. If the average score on the exam is 80 points with a standard deviation of 12 points what is the minimum score that an applicant must make to be considered eligible?

****

Normal Distribution

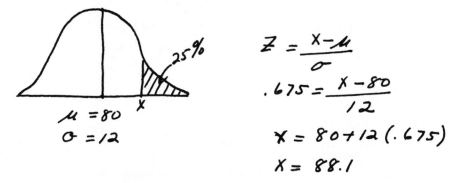

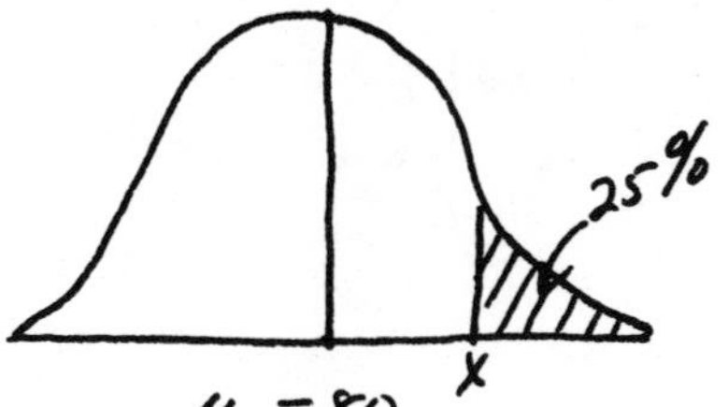

$\mu = 80$

$\sigma = 12$

$$Z = \frac{X-\mu}{\sigma}$$

$$.675 = \frac{X-80}{12}$$

$$X = 80 + 12(.675)$$

$$X = 88.1$$

6-12

Find the largest error one can expect to make with a probability of 0.90 when using the mean of a random sample of size n=100 to estimate the mean of a population having a variance of 2.56.

****

$$P\left\{ Z \leq \left| \frac{\bar{x}-\mu}{\sigma/\sqrt{n}} \right| \right\} = 0.90$$

$$\sigma = \sqrt{2.56} = 1.60$$

$$\therefore Z_{\alpha=0.10} = 1.28$$

$$\text{ERROR } |\bar{x}-\mu| = 1.28(1.60)/\sqrt{100}$$

$$= 0.2048$$

6-13

The height of trucks on an interstate is approximately Normally distributed with mean 10 ft and standard deviation 1.5 ft. Find the clearance, D, at an overpass if the probability that a truck will clear it is 0.999.

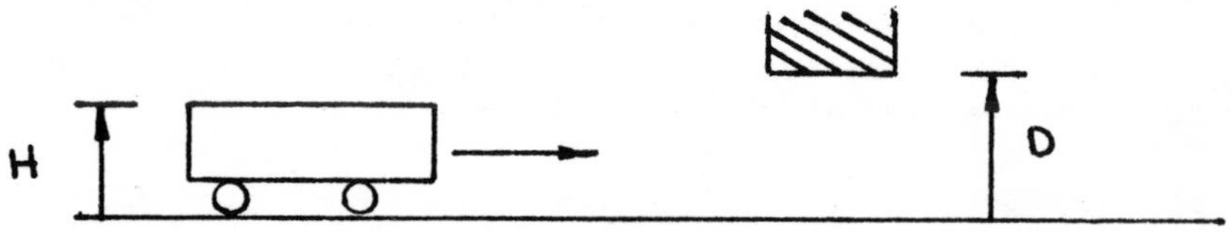

****

$H \sim N(\mu, \sigma)$ where $\mu = 10$ ft , $\sigma = 1.5$ ft

$$\Pr(H < D) = \Pr\left[\frac{H-10}{1.5} < \frac{D-10}{1.5}\right] = \Pr\left[Z < \frac{D-10}{1.5}\right]$$

$$= 0.999$$

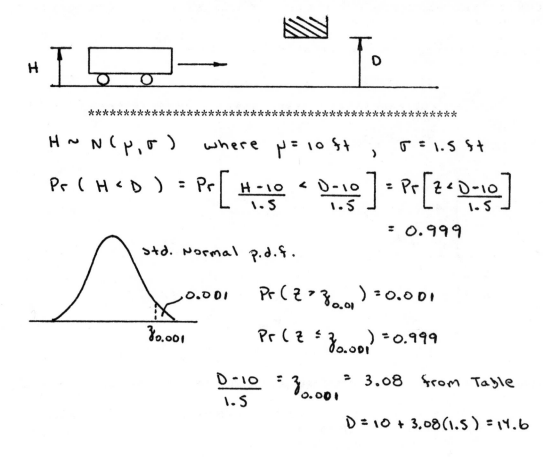

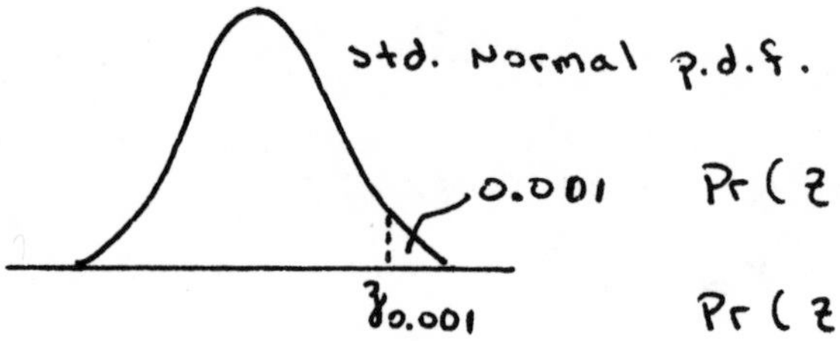

$$\Pr(Z > z_{0.01}) = 0.001$$

$$\Pr(Z \le z_{0.001}) = 0.999$$

$$\frac{D-10}{1.5} = z_{0.001} = 3.08 \text{ from Table}$$

$$D = 10 + 3.08(1.5) = 14.6$$

6-14

Find the size of sample necessary to estimate, at the one percent level of significance, the mean strength of concrete within 25 psi if the standard deviation is 100 psi. The strength is normally distributed.

****

$$Z = \frac{\bar{x} - \mu}{\sigma/\sqrt{n}} \qquad 2.575 = \frac{25 \text{ psi}}{100 \text{ psi}/\sqrt{n}}$$

$$\sqrt{n} = (4)(2.575) = 10.3$$

$$n = 10.3^2 = 106.09$$

Use 107 Samples

6-15

The average life of a certain type of compressor is 10 years with a standard deviation of 1 year. The manufacturer replaces free all compressors that fail while under guarantee. If they are willing to replace only 3% of all the compressors sold, how long a guarantee should they offer? Assume the lives of the compressors follow a Normal distribution.

****

Let T = Life time of compressor, $T \sim N(\mu, \sigma)$

$\mu = 10$ yr

If G is the guarantee period, then $\sigma = 1$ yr

$Pr(T < G) = 0.03$

$$Pr\left[\frac{T-\mu}{\sigma} < \frac{G-\mu}{\sigma}\right] = Pr\left[Z < \frac{G-10}{1}\right] = 0.03$$

Since $Pr(Z < z_{0.97}) = 0.03$

it follows that

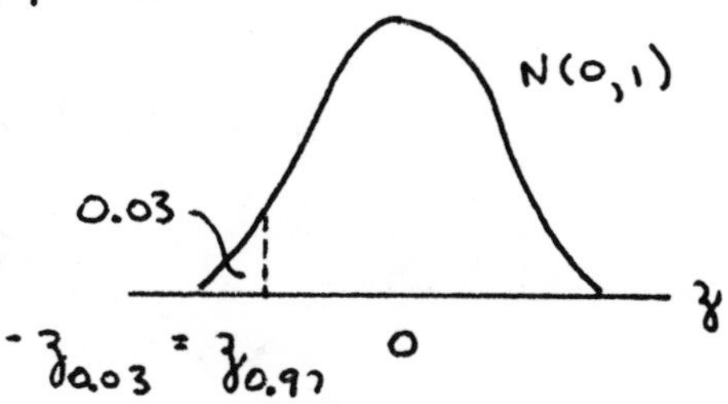

$$G - 10 = z_{0.97} = -z_{0.03}$$

$$G = 10 - z_{0.03} = 10 - 1.88 = 8.12 \text{ yr}$$

6-16

Given that for a standard normal distribution, P(Z > 1)=.1587 find:

P(-1 < Z < 0), P(-1 < Z <1), P(Z <1).

****

$$P(-1 < Z < 0) = P(Z < 1) - P(Z > 0)$$
$$= (1 - .1587) - .5 = 34.13\%$$
$$P(-1 < Z < 1) = 2P(-1 < Z < 0) = 68.26\%$$
$$P(Z < 1) = 1 - P(Z > 1) = (1 - .1587) = 84.13\%$$

6-17

Find the maximum likelihood estimator for the "t" based on n observations for the following distributions:

(a) $f(x_i;t) = \frac{t}{(2\Pi)^{1/2}} e^{-(tx_i)^2/2}$, $t > 0$, $-\infty < x < \infty$

(b) $f(x_i;t) = (1 + 2t)x_i^{2t}$, $t > 0$, $0 < x_i < 1$.

****

(a) $$L = \prod_{i=1}^{n} f(x_i;t) = (t/\sqrt{2\pi})^n e^{-\frac{t^2}{2}\sum_{i=1}^{n} x_i^2}$$

$$\mathcal{L} = \ln L = n \ln t - \frac{n}{2}\ln(2\pi) - \frac{t^2}{2}\sum_{i=1}^{n} x_i^2$$

$$\frac{\partial \mathcal{L}}{\partial t} = \frac{n}{t} - \hat{t}\sum_{i=1}^{n} x_i^2 = 0$$

$$\therefore \hat{t}^2 = n/\sum_{i=1}^{n} x_i^2, \quad \hat{t} = \sqrt{n/\sum_{i=1}^{n} x_i^2}$$

(b) $$L = \prod_{i=1}^{n} f(x_i;t) = \prod_{i=1}^{n} (1+2t)\, x_i^{2t}$$

$$\mathcal{L} = \ln L = n \ln(1+2t) + \sum_{i=1}^{n} 2t \ln x_i^2$$

$$\frac{\partial \mathcal{L}}{\partial t} = \frac{2n}{1+2\hat{t}} + 2\sum_{i=1}^{n} \ln x_i^2 = 0$$

$$\therefore 1+2\hat{t} = -n/\sum_{i=1}^{n} \ln x_i^2$$

OR $$\hat{t} = -\frac{1}{2}\left[1 + n/\sum_{i=1}^{n} \ln x_i^2\right]$$

6-18

The ounces of fluid dispensed from a soft drink machine is Normally distributed with mean μ = 9 ounces and standard deviation σ unknown. Find σ if there is a 2.5% chance of overfilling a 10 ounce cup.

****

Let r.v. Y = Ounces of fluid dispensed

$Y \sim N(\mu, \sigma)$ where $\mu = 9$ and σ is unknown

$Pr(Y > 10) = 0.025$

$$Pr\left[\frac{Y-\mu}{\sigma} > \frac{10-\mu}{\sigma}\right] = 0.025$$

$$Pr\left[Z > \frac{10-9}{\sigma}\right] = 0.025$$

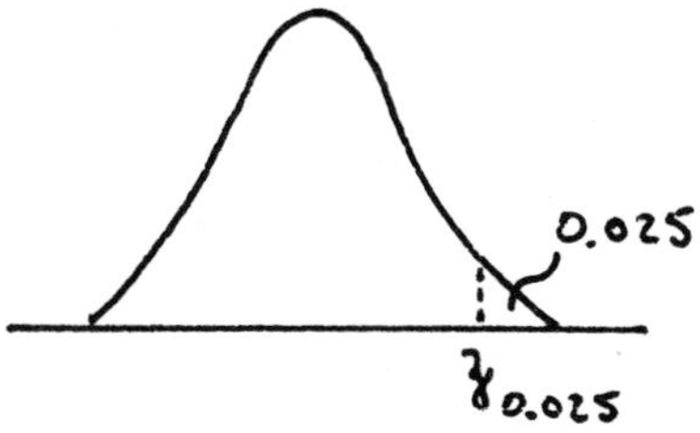

Also, $Pr(Z > z_{0.025}) = 0.025$

Therefore, $\frac{1}{\sigma} = z_{0.025} = 1.96$ from Table

$$\sigma = \frac{1}{1.96} = 0.51 \text{ ounces}$$

6-19

Determine the maximum allowable standard deviation for a plastic production process with a mean strength of 32.0 kN/sq m. The minimum allowable design strength is 30.0 kN/sq m. Specifications allow one sample in 100 to fail (to be below design strength). The strength is normally distributed.

****

$F(z) = 0.49$ $\quad Z = -2.327$

$z = \frac{x-\mu}{\sigma}$ $\quad -2.327 = \frac{30-32}{\sigma}$ $\quad \sigma = 0.86$ kN/sq. m.

6-20

The grades on an examination are Normally distributed with mean μ and standard deviation σ. The minimum grade for an A is 92 and the minimum grade to pass (not get an F) is 57. Find μ and σ if 15% of the class get A's and 10% get F's.

****

Let S = Score on Exam , $S \sim N(\mu, \sigma)$

$\Pr(S \geq 92) = 0.15$

$\Pr(S < 57) = 0.10$

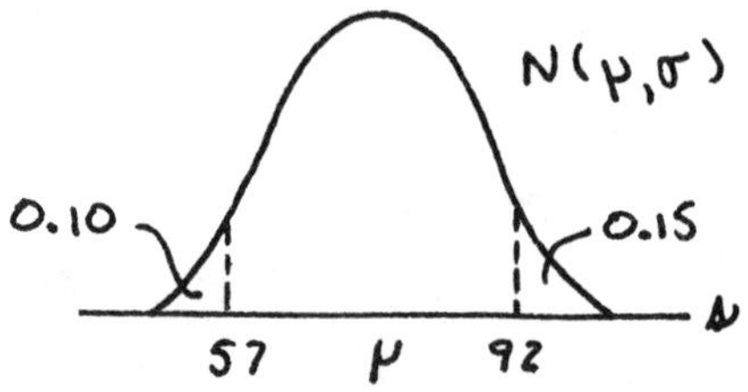

Standardizing the above

$$\Pr\left[Z \geq \frac{92-\mu}{\sigma}\right] = 0.15$$

$$\Pr\left[Z < \frac{57-\mu}{\sigma}\right] = 0.10$$

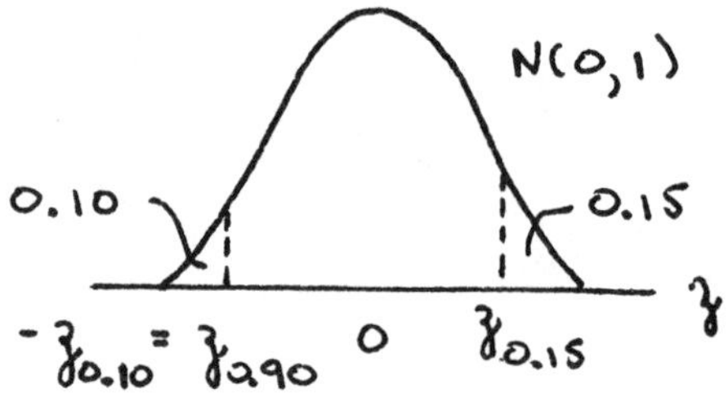

It follows that

$$\frac{92-\mu}{\sigma} = z_{0.15} = 1.04$$

$$\frac{57-\mu}{\sigma} = -z_{0.10} = -1.28$$

Solving for μ and σ yields $\mu = 76.3$ & $\sigma = 15.1$

6-21

A manufacturer of car batteries desires to establish a warranty period so that less than 12% of batteries would fail before the warranty expires. The only data available from the manufacturer are the following records of battery lives in years (recorded in the order reading across). The population was determined to be normally distributed based on the data, and range and mean control charts showed the data to be in control. Determine a warranty period in months for the batteries to satisfy the needs of the manufacturer.

Lives (years):							
2.2	4.1	3.5	4.5	3.2	3.7	3.0	2.6
3.4	1.6	3.1	3.3	3.8	3.1	4.7	3.7
2.5	4.3	3.4	3.6	2.9	3.3	3.9	3.1
3.3	3.1	3.7	4.4	3.2	4.1	1.9	3.4
4.7	3.8	3.2	2.6	3.9	3.0	4.2	3.5

****

Mean life $= \mu \cong \frac{\Sigma \text{ lives}}{40} = \frac{136.5}{40} = 3.41$

Standard deviation $= \sigma \cong \sqrt{\frac{(40)(\Sigma \text{ lives}^2) - (\Sigma \text{ lives})^2}{(40)(39)}}$ * $= 0.703$

Based on the normal distribution shown, x (the desired warranty life in years) is determined as follows using a normal table, where Φ represents the area under the normal curve from $-\infty$ to the value desired:

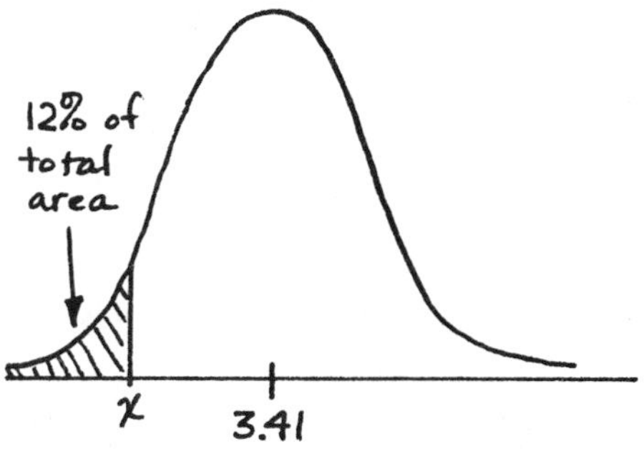

$$\Phi\left(\frac{x - 3.41}{0.703}\right) = 0.12$$

$\frac{x - 3.41}{0.703} = -1.175$ (the value in the normal table to the left of which is 12% of the area under the curve of the standard normal)

Solving for x, $x \cong 2.59$ yrs. or 31 months for the warranty period

* Σ lives $= 136.5$, $\Sigma \text{ lives}^2 = 485.07$

6-22

The travel time experienced by a driver going from Austin to Houston along U.S. Highway 290 can be viewed as a normally distributed random variable with a mean of 3.50 hours and a standard deviation of 0.45 hours.

a) Calculate the probability that a randomly selected driver experiences a travel time in excess of 4.0 hours.

b) Consider an experiment involving the measurement of the travel time of 50 different randomly and independently selected drivers going from Austin to Houston along U.S. Hishway 290.

   i) Calculate the probability that the average travel time (taken over the 50 observations) is in excess of 4.0 hours.

   ii) What is the expected number of drivers in this sample who will have experienced a travel time in excess of 4.0 hours? What is its standard deviation?

c) Some drivers, after arriving in Houston, may decide to continue on to Galveston. We can assume that the travel time from Houston to Galveston is normally distributed with a mean of 1.25 hours and a standard deviation of 0.75 hours. We can further assume that the travel time between Austin and Houston is independent from that between Houston and Galveston.

   For a randomly selected driver going from Austin to Houston and then continuing on to Galveston, calculate the probability that the total travel time (assuming no stops along the way) is in excess of 5 hours but no more than 6 hours.

****

a) Let X denote the travel time between Austin and Houston.

$$X \sim N(3.5,\ (0.45)^2)$$

$$p(X>4) = 1 - p(X \le 4) = 1 - p\left(Z \le \frac{4-3.5}{0.45}\right)$$

$$= 1 - \Phi(1.111) = 1 - 0.8667 = \underline{\underline{0.1333}}$$

[note: the function $\Phi(\cdot)$ is the cumulative standard normal distribution, which is usually tabulated].

b) Consider the random variable $\overline{X}$, which denotes the mean of a sample of 50 travel time observations.

$$\overline{X} \sim N\left(3.5, \frac{(0.45)^2}{50}\right)$$

i) $p(\overline{X} > 4.0) = 1 - p(\overline{X} \leq 4.0) = 1 - p\left(Z \leq \frac{4.0 - 3.5}{0.45/\sqrt{50}}\right)$

$= 1 - \Phi(7.856) \simeq 1 - 1 = 0$

ii) Let N denote the number of drivers, out of 50, who experience a travel time in excess of 4 hours.

Each driver can be viewed as an independent Bernoulli trial (with "success" $\Leftrightarrow$ "travel time >4 hours").

The probability of success is constant across trials, and was found in part (a) to be equal to 0.1333.

N is therefore a binomially distributed random variable, with $n = 50$ and $p = 0.1333$.

$E[N] = n \cdot p = 50 \times 0.1333 = 6.665$ drivers

$V[N] = np(1-p) = 50 \times 0.1333 \times (1 - 0.1333)$

$= 5.777 \Rightarrow$ std. dev. $= \sqrt{V[N]} = 2.404$ drivers.

c) Let Y denote the travel time between Houston and Galveston, and W the total travel time between Austin and Galveston via Houston. $Y \sim N(1.25, (0.75)^2)$

$W = X + Y$; since both X and Y are normally distributed,

$W \sim N([3.5 + 1.25], [0.45^2 + 0.75^2])$, i.e. $W \sim N(4.75, 0.765)$

$$p(5 \leq W \leq 6) = p\left(\frac{5 - 4.75}{\sqrt{0.765}} \leq Z \leq \frac{6 - 4.75}{\sqrt{0.765}}\right)$$

$= \Phi(1.429) - \Phi(0.286) = 0.9235 - 0.6126 = \underline{0.3109}$

6-23

A random variable X is normally distributed with mean, $\mu = 40$ and variance, $\sigma^2 = 16$. Determine the probability, $P(X > 50)$.

****

$$P(X > 50) = 1 - P(X \leq 50)$$

$$= 1 - P\left(\frac{X-\mu}{\sigma} \leq \frac{50-40}{\sqrt{16}}\right)$$

$$= 1 - P(Z \leq 2.5) = 1 - \Phi(2.5)$$

where Z is a standard normal variable and Φ is a standard normal distribution function. Selecting the correct value of the distribution function from the appropriate standard normal probability tables, one has

$$P(X > 50) = 1 - 0.9938 = 0.0062$$

$$\simeq 0.6\,\%$$

6-24

A pelletizing process is said to be in control if the mean crushing strength is 250 kilograms. It is known that the crushing strength measurements are normally distributed with standard deviation of 40 kilograms. Periodic random samples of size 10 are taken from this process and the process is said to be "out of control" if a sample mean is less than 230.0 kg (crushing strength is measured to the nearest tenth of a kilogram).

a) Find the probability that, if the process is under control, a randomly selected sample will indicate that it is not ______________.

b) Find the probability that, in a sample, at least three pellets will have a crushing strength less than 216.4 kg ______________________.

c) If a sample yields a mean crushing strength over 275.0 kilograms, would you say that the process is in control? Explain (brief, please)

****

$X \sim N(\mu = 250, \sigma = 40)$

a) $Pr(\text{out of control}) = Pr(\bar{X} < 230.0)$

$$= Pr\left(z < \frac{230.0 - 250.0}{40/\sqrt{10}}\right) = Pr(z < -1.58)$$

$$= 1 - Pr(z < 1.58) = 1 - 0.9429 = 0.0571$$

b) $Pr(\text{one pellet showing crushing strength} < 216.4)$

$$= Pr\left(z < \frac{216.4 - 250}{40}\right) = Pr(z < -0.84) = 0.20$$

$$Pr(X \geq 3;\ n = 10,\ p = 0.20) = 1 - B(2;\ 10, 0.20)$$

$$= 0.3272$$

c) $$Pr(\bar{x} > 275) = Pr\left(z > \frac{275 - 250}{40/\sqrt{10}}\right)$$

$$= Pr(z > 1.98) \simeq 0.024$$

PROB. TOO SMALL ∴ OUT OF CONTROL!

6-25

Peter Piper the Pickle Packer provides a written guarantee with each bottle of his pickles that every pickle will be at least 8 cm. long. His bottles are such that the longest pickle that they can hold is 10.5 cm. If Peter Piper picks a peck of pickles whose lengths are normally distributed with a mean of 9 cm and a standard deviation of .8 cm, what percent of the pickles from the peck that Peter Piper picked will have to be discarded because they either don't meet his guaranteed standard or they wont fit in the bottle?

How tall would the bottles have to be if Peter wishes to discard no more than 5 percent of the pickles he picks from from the peck?

****

Normal Distribution

$\mu = 9 \quad \sigma = .8$

$Z = \frac{X-\mu}{\sigma} = \frac{8-9}{.8} = -1.25$

$P(X < 8) = .1055$

$Z = \frac{X-\mu}{\sigma} = \frac{10.5-9}{.8} = 1.87$

$P(X > 10.5) = .0307$

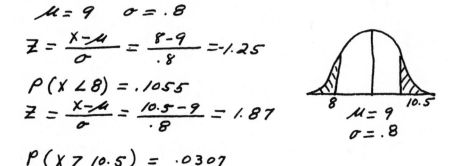

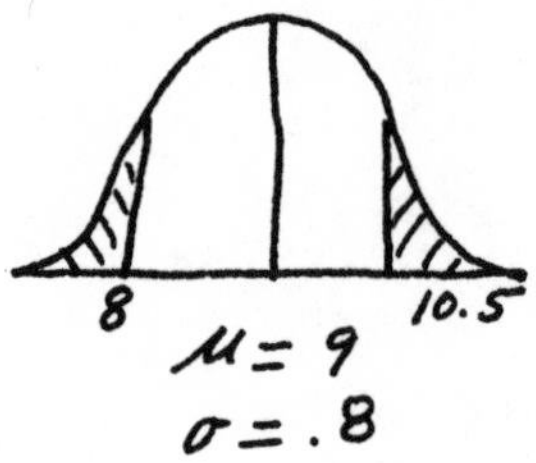

a) $P(8 > X > 10.5) = .1055 + .0307 = .1316 = 13.16\%$

b) It will be impossible to reduce the total rejects to 5% by increasing the bottle length since 10.5% is dicarded because they are too short.

NORMAL APPROXIMATION TO BINOMIAL

6-26

The probability of successfully rehabilitating an alcoholic is 0.25. How many alcoholics must be treated so that at least 40 are successfully rehabilitated with a probability of 0.95?

****

S = SUCCESS, F = FAILURE, N = # SUCCESSES

$Pr\{S\} = 0.25$, $Pr\{F\} = 0.75$

N IS A BINOMIAL RANDOM VARIABLE WITH PARAMETERS OF n (THE UNKNOWN) AND $p = 0.25$.

$$E[N] = np = 0.25n, \quad VAR[N] = npq = 0.25 \times 0.75n$$

$$Pr\{N > 40\} = 0.95$$

$$Pr\{N > 40\} = Pr\left\{ \frac{N - .25n}{\sqrt{.25 \times .75n}} > \frac{40 - .25n}{\sqrt{.25 \times .75n}} \right\}$$

$$\cong Pr\{Z > z_a\} = 0.95 \quad \text{(NORMAL APPROXIMATION)}$$

$$z_a = -1.645 = \frac{40 - .25n}{\sqrt{.1875n}}, \quad \text{LET } \sqrt{n} = x$$

$$.25x^2 - .7123x - 40 = 0$$

USING THE QUADRATIC FORMULA, THERE IS ONLY ONE POSITIVE ROOT, i.e.,

$$x = 14.15 \quad \text{OR} \quad n \cong 200.$$

6-27

Sixty percent of all registered voters are Democrats and forty percent are Republicans. In a sample of 150 randomly selected voters, what is the probability that at least half of them will be Republicans?

****

FOR PARENT BINOMIAL: USING X = NO. OF REPUBLICANS

$p = .4\ ,\ n = 150;\ \mu = np = 60$

$\sigma = \sqrt{npq} = \sqrt{60(.6)} = 6$

USING NORMAL APPROX:

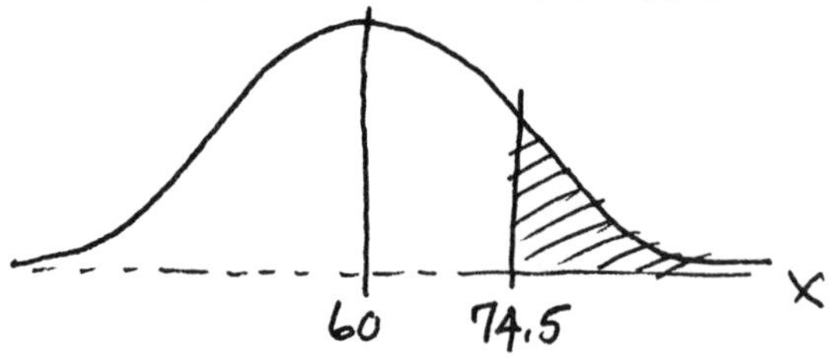

$Z = \dfrac{74.5 - 60}{6} = 2.42$

AT LEAST HALF IS $X \geq 75$; CONTINUITY CORR. MAKES $X \geq 74.5$

$P(X \geq 75) = P(Z > 2.42) = .0078$

6-28

A grading system generally used is 90 or above A; 80-90 B; 70-80 C; 60-70 D; below 60 F. If the distribution of grades is approximately normally distributed and 5% of the grades are A and 5% are F, a) find the mean numerical grade, b) find the standard deviation of the numerical grade, c) find % of the grades that will be C.

****

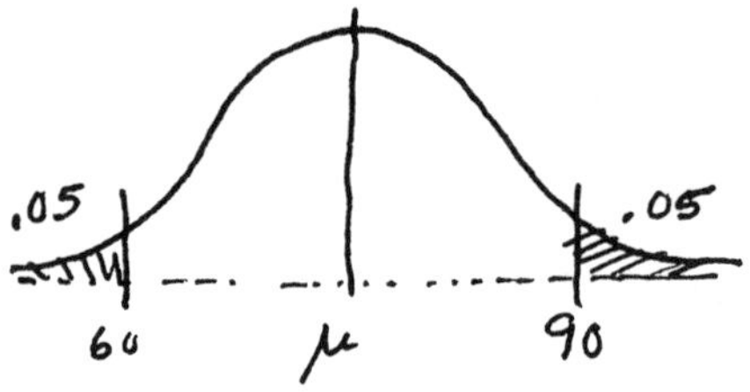

(a) BY SYMMETRY, $\mu = 75$

(b) FOR A .05 tail area, $Z = 1.645$

$\therefore \dfrac{90 - 75}{\sigma} = 1.645$ and $\sigma = 9.12$

(c) $Z = \dfrac{80 - 75}{9.12} = 0.548$

$\therefore$ Area between μ and 80 is $.7081 - .5 = .2081$

and $P(70 < X < 80) = 2(.2081) = .4162$ or 41.6%

6-29

The reliability of each of 50 independent components in a system is equal to 0.90 . How confident can the manufacturer be that at least 42 of the components function properly? Use the normal distribution to develop a confidence level.

****

This is actually a binomial problem (a component either functions properly or does not) for which the normal distribution will give an approximation if a correction for continuity is made (ie., a discrete distribution is being approximated by a continuous one). For the binomial,

$$\text{Mean} = \mu = (\text{sample size } n) \times (\text{reliability } p) = (50)(0.90) = 45$$

$$\text{Variance} = \sigma^2 = np(1-p) = (50)(0.90)(1-0.90) = 4.5$$

Let the random variable x count the number of components functioning properly in a sample of 50. The desired probability that at least 42 of the 50 function properly is

$P(x \geq 42) \cong P(41.5 \leq x \leq 50.5)$ using the standard continuity correction (see figure).

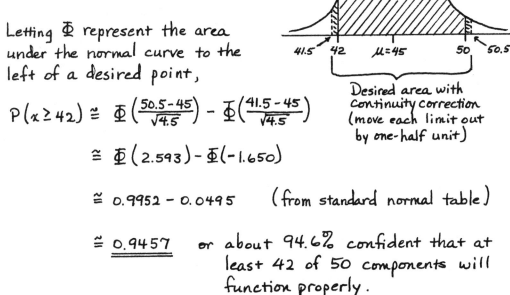

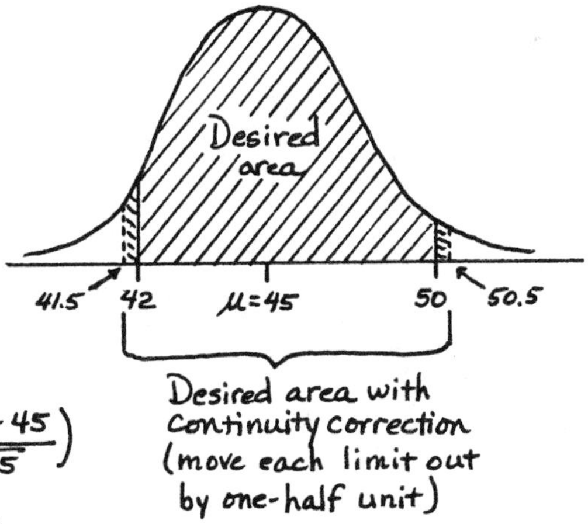

Letting Φ represent the area under the normal curve to the left of a desired point,

$$P(x \geq 42) \cong \Phi\left(\frac{50.5-45}{\sqrt{4.5}}\right) - \Phi\left(\frac{41.5-45}{\sqrt{4.5}}\right)$$

$$\cong \Phi(2.593) - \Phi(-1.650)$$

$$\cong 0.9952 - 0.0495 \quad \text{(from standard normal table)}$$

$$\cong \underline{\underline{0.9457}}$$

or about 94.6% confident that at least 42 of 50 components will function properly.

6-30

Joe and Bob flip a fair coin every morning to see who buys the coffee. Find the probability that Bob will buy coffee more than 140 days out of 250 working days in a year.

****

Let r.v. X = Number of days Bob buys coffee in a year

and $p = Pr$ (Bob buys coffee on a given day)

Then X is binomially distributed, $X \sim b(n,p)$
$n = 250$ & $p = 0.5$

$$Pr(X > 140) = 1 - Pr(X \le 140)$$

$$= 1 - \sum_{x=0}^{140} b(x; 250, 0.5)$$

$$= 1 - B(140; 250, 0.5)$$

Since the tables of $B(x; n, p)$ generally do not go as high as $n = 250$, the Normal distribution can be used to give an approximate solution.

Using $X \sim N(\mu, \sigma^2)$ with $\mu = np = 250(0.5) = 125$
$\sigma^2 = np(1-p) = 125(0.5) = 62.5$

Including the continuity correction,

$$Pr(-0.5 \le X \le 140.5) = Pr\left[\frac{-0.5 - 125}{\sqrt{62.5}} \le Z \le \frac{140.5 - 125}{\sqrt{62.5}}\right]$$

$$= Pr(-15.9 \le Z \le 1.96)$$

$$= F_Z(1.96) = 0.975$$

$$Pr(X > 140) = 1 - 0.975$$

$$= 0.025$$

6-31

In tossing a coin find (a) the probability that in 8 tosses the first five will be heads & the last three tails, (b) the probability that in 8 tosses exactly half are heads, (c) by an approximation the probability that less than 240 heads will occur in 500 tosses? OR less than 190 in 400 tosses?

****

(a) FOR A PARTICULAR SEQUENCE OF HEADS & TAILS: P IS $\left(\frac{1}{2}\right)^8 = \frac{1}{256}$

(b) BY BINOMIAL PROBS: P IS $\binom{8}{4}\left(\frac{1}{2}\right)^8 = \frac{70}{256}$

(c) BY NORMAL APPROX: LESS THAN 240 IN 500 TOSSES:

$\mu = np = 500 \cdot \frac{1}{2} = 250$

$\sigma^2 = npq = 250 \cdot \frac{1}{2} = 125$; $\sigma = \sqrt{125} = 11.18$

USING CONTINUITY CORRECTION: $Z = \dfrac{239.5 - 250}{11.18} = -.94$

$\therefore$ P(LESS THAN 240) $\cong$ AREA UNDER NORMAL CURVE BELOW $Z = -.94$

$= .1736$

LIKEWISE, LESS THAN 190 IN 400 TOSSES:

$\mu = np = 200$ $\sigma^2 = npq = 100$; $\sigma = 10$

$Z = \dfrac{189.5 - 200}{10} = -1.05$

PROB. IS (AREA BELOW $Z = -1.05$) $= .1469$

6-32

A sample of 50 electric bulbs was picked, one bulb at a time, from a large batch. If the probability of picking a good bulb in each drawing is 0.9, what is the probability that the number of bad bulbs in the sample is less than three? Determine the percentage error introduced if the normal distribution approximation is used to solve this binomial distribution problem.

****

Binomial probability distribution is given by

$$p(n) = {}^{m}C_{n}\, p^{n}(1-p)^{m-n}$$

$$n = 0, 1, 2, \cdots, m$$

with mean $\mu = mp$ and variance $\sigma^2 = mp(1-p)$.

Random variable N = number of bad bulbs in a sample of 50. Using binomial distribution for N with $m = 50$ and p = probability of picking a bad bulb $= 1 - 0.9 = 0.1$

$$P(N=0) = p(0) = {}^{50}C_{0}(0.1)^{0} \times 0.9^{50} = (0.9)^{50} = 0.00515$$

$$P(N=1) = p(1) = {}^{50}C_{1}(0.1) \times 0.9^{49} = 0.02863$$

$$P(N=2) = p(2) = {}^{50}C_{2}(0.1)^{2} \times 0.9^{48} = 0.07794$$

$\therefore$ Probability of having less than 3 bad bulbs

$$= p(0) + p(1) + p(2) = \underline{\underline{0.11172}}$$

$$\mu = 50 \times 0.1 = 5 \quad \text{and} \quad \sigma = \sqrt{50 \times 0.1 \times 0.9} = 2.1213$$

Since N is a "discrete" random variable (that takes values 0, 1, 2, 3, $\cdots$) we must use a <u>continuity correction</u> when using the normal-distribution approximation to represent N. Thus use $P(N < 2.5)$ in place of $P(N < 3)$.

Define the normalized random variable

$Z = \frac{N-\mu}{\sigma}$ having zero mean and unity std. deviation.

$P(N<2.5) = P\left(Z < \frac{2.5-\mu}{\sigma}\right) = P\left(Z < \frac{2.5-5}{2.1213}\right)$

$= P(Z < -1.1785)$

$= 1 - 0.8807$ (From tables)

$= 0.1193$

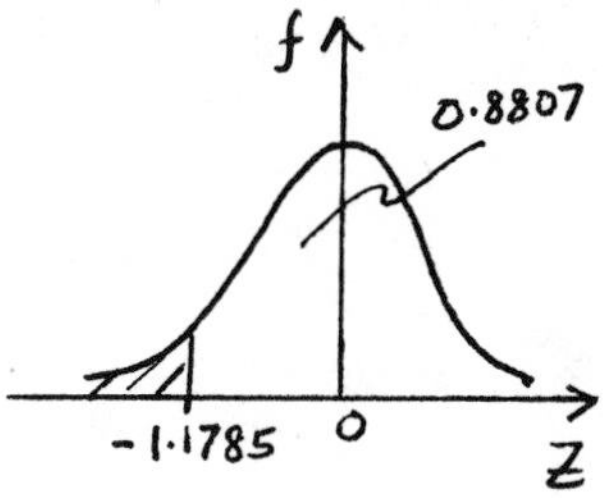

Percentage error $= \frac{(0.1193 - 0.11172)}{0.11172} 100\ \%$

$= \underline{\underline{6.8\ \%}}$ which is acceptable.

6-33

In 144 throws of an unbiased coin, we would expect that at least 90% of the time the number of heads would fall between what limits?

****

DISTRIBUTION IS BINOMIAL: $n = 144$, $p = \frac{1}{2}$ $\mu = np = 72$, $\sigma = \sqrt{npq} = 6$

USING NORMAL APPROX.

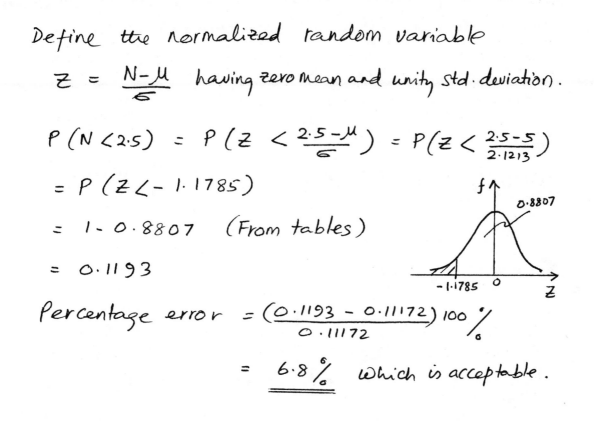

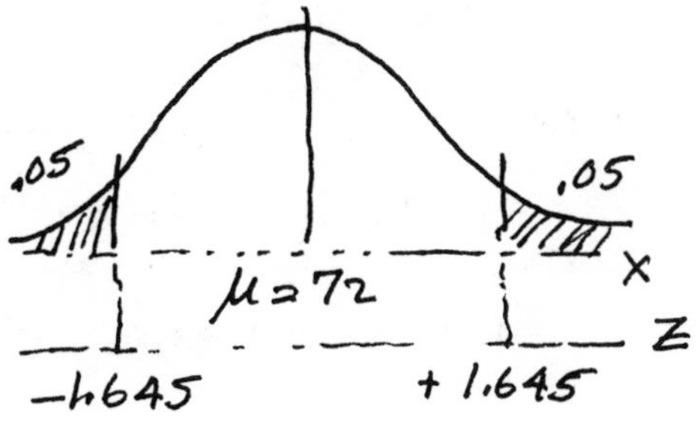

$X = 72 \pm 1.645(6)$

$= 72 \pm 9.87$ → 81.87, 62.13

CHANGING TO INTEGERS:

$63 \leq X \leq 81$

6-34

The probability a unit produced by a machine turns out to be defective is $p = 1/100$.

a. A random sample of 10 units produced by this machine is selected. Find P (1 unit is defective) ________.

b. A random sample of 300 units is inspected. Find P (at least 4 are defective).

c. If, due to a malfunction, the probability of a defective unit becomes $p = 0.175$, what's the probability that, among 200 units randomly inspected from a batch produced by the malfunctioning machine, at most 30 will be defective ________.

****

BINOMIAL SITUATION : $p = 0.01$

a) $b(1; 10, 0.01) = \binom{10}{1}(0.01)^1(0.99)^9 = 0.091$

b) $n > 20$, p small : Use Poisson approximation

with $\lambda = np = (300)(0.01) = 3$

$Pr(X \geq 4, \lambda = 3) = 1 - Pr(X \leq 3, \lambda = 3)$

$= 1 - 0.647 = 0.353$

c) $np = (200)(0.175) = 35$ Use Normal approx.

$\sigma = \sqrt{np(1-p)} = \sqrt{(200)(0.175)(0.825)} = 5.374$

$Pr(X \leq 30) = Pr\left(z < \frac{30.5 - 35}{5.374}\right)$ ← continuity correction

$= Pr(z < -0.84) = 1 - Pr(z < 0.84)$

$= 1 - 0.7995 = 0.2005$

6-35

Two friends, Wayne and Henry, worked together for the summer and played 120 sets of tennis during that time. Wayne won 70 of these sets, while Henry only won 50. If they are equally skilled at tennis, what is the probability that Wayne would win 70 or more? What conclusion might you draw from this result?

****

USING THE NORMAL APPROXIMATION TO THE BINOMIAL (WHICH IS PERMISSABLE BECAUSE N IS LARGE AND P IS CLOSE TO 0.5, IN THIS CASE P = 0.5),

$$P[X \geq 70 \mid p = 0.5] = P\left[Z > \frac{69.5 - 60}{5.48}\right]$$

$$= P[Z > 1.73] = 1 - 0.9582 = 0.0418$$

SINCE THE PROBABILITY OF 70 OR MORE WINS IS ONLY 4.2%, IT DOES NOT SEEM LIKELY THAT HENRY AND WAYNE ARE EQUALLY SKILLED PLAYERS.

6-36

A consumer has found that 20 percent of the eggs she buys from a certain market have double yokes. She has just purchased 3 dozen eggs from the market. What is the probability that her purchase will contain at least 10 double yoke eggs?

****

Binomial Probability Distribution

$p = .2 \quad q = .8 \quad n = 36 \quad P(X \geq 10) = ?$

However since $np > 5$ and $nq > 5$ the normal approximation to the binomial is appropriate

$\mu = np = 36(.2) = 7.2 \qquad \sigma = \sqrt{npq} = 2.4$

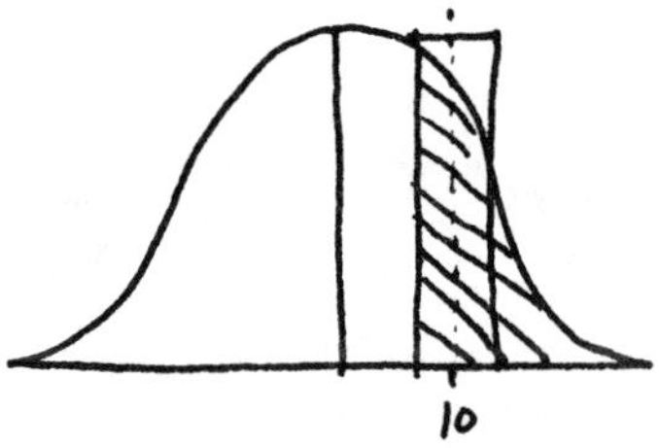

$$Z = \frac{X-\mu}{\sigma} = \frac{9.5-7.2}{2.4} = .917$$

from normal table $P(X \geq 10) = 18\%$

LOG - NORMAL

6-37

Maximum flood flows past a bridge during the past ten years are summarized below

Year	1	2	3	4	5	6	7	8	9	10
Flow, 1000 cfs	23	38	17	210	62	142	43	29	71	31

Estimate the flood flow that would be expected to be exceeded once in 100 years if the flood flows are assumed to follow a log normal distribution.

****

Year	Flow, 1000 cfs	$\ln Q = X$	X^2
1	23	3.13	9.83
2	38	3.64	13.25
3	17	2.83	8.01
4	210	5.35	28.62
5	62	4.13	17.06
6	142	4.96	24.60
7	43	3.76	14.14
8	29	3.37	11.36
9	71	4.26	18.15
10	31	3.43	11.76
	Totals	38.86	156.78

$\bar{X} = \frac{\Sigma X}{n} = \frac{38.86}{10} = 3.886$ $\quad F(z) = 0.49 \quad z = 2.327$

$s^2 = \frac{\Sigma X^2 - (\Sigma X)^2/n}{n-1} = \frac{156.78 - (38.86)^2/10}{10-1} = \frac{5.77}{9} = 0.64$

$s = 0.80 \quad z = \frac{X - \bar{X}}{s} \quad 2.327 = \frac{X - 3.886}{0.80} \quad X = 5.748$

$e^{5.748} = 313 \quad Q = \underline{313,000 \text{ cfs}}$

6-38

Find the probability that random variable X is between 10 and 20 if ln X is Normally distributed with mean 3 and standard deviation 2.

****

$\ln X \sim N(\mu, \sigma)$ where $\mu = 3$, $\sigma = 2$

$$\begin{aligned}
\Pr(10 \le X \le 20) &= \Pr(\ln 10 \le \ln X \le \ln 20) \\
&= \Pr\left[\frac{\ln 10 - 3}{2} \le \frac{\ln X - \mu}{\sigma} \le \frac{\ln 20 - 3}{2}\right] \\
&= \Pr(-0.3487 \le z \le -0.0021) \\
&= F_z(-0.0021) - F_z(-0.3487) \\
&= 0.5 - [1 - F_z(0.3487)] \\
&= -0.5 + 0.636 \\
&= 0.136
\end{aligned}$$

EXPONENTIAL

6-39

The number of passengers arriving at a particular bus stop is Poisson distributed with a mean rate of 1.2 passengers per minute. Calculate the probability that the time between two consecutive passenger arrivals exceeds 2 minutes. What is the distribution of the time between consecutive arrivals?

****

Let T denote the time between consecutive arrivals.

Since arrivals follow a Poisson process, T follows the exponential distribution, with pdf $f_T(t) = 1.2e^{-1.2t}$, for $t \geqslant 0$

proba (time between consecutive arrivals exceeds 2 minutes)

$$= p(T \geqslant 2) = 1 - F_T(t),$$

where $F_T(t)$ is the exponential cdf, given by:

$$F_T(t) = 1 - e^{-1.2t}$$

thus $p(T \geqslant 2) = 1 - (1 - e^{-1.2\times 2}) = e^{-1.2\times 2}$

$$= e^{-2.4} = \underline{\underline{0.091}}.$$

Alternatively,

if N denotes the number of Poisson arrivals in a 2-minute interval,

$$p(T \geqslant 2) = p(N = 0) = \frac{(2\times 1.2)^0 e^{-2\times 1.2}}{0!} = e^{-2.4} = 0.091.$$

6-40

A CHECKOUT COUNTER AT A SUPERMARKET CAN HANDLE CUSTOMERS AT A RATE OF 20 PER HOUR.

WHAT IS THE PROBABILITY THAT A CUSTOMER WILL TAKE OVER 5 MINUTES TO BE CHECKED OUT ?

****

THE CHECKOUT PROCESS CAN BE MODELED WITH A NEGATIVE EXPONENTIAL DISTRIBUTION.

$$\lambda = 20/\text{hr} \qquad \Theta = \frac{1}{\lambda} = \frac{1}{20}\frac{\text{hr}}{\text{CUST}} = 3\text{ MIN}$$

$$P = (t > 5) = 1 - P(t < 5)$$

$$= 1 - \left[1 - e^{-\frac{t}{\theta}}\right] = e^{-\frac{t}{\theta}}$$

$$= e^{-\frac{5}{3}} = 0.189$$

6-41

If the number of customers arriving per time unit at a grocery store is a random variable having a Poisson distribution, the time between arrivals is a random variable having the exponential distribution. If customers arrive in a Poisson fashion, at an average rate of 30 per hour, what is the probability that the time to the next arrival will be more than three minutes?

****

AVERAGE TIME BETWEEN ARRIVALS = 2 MIN.

$$P[X > 3] = \int_3^{\infty} \tfrac{1}{2} e^{-\frac{1}{2}x}\,dx = -e^{-\frac{1}{2}x}\Big]_3^{\infty}$$

$$= e^{-3/2} = 0.2231$$

6-42

The lifetime of a certain kind of battery is a random variable which has an exponential distribution with θ = 250 hours. What is the probability that such a battery will last at most 200 hrs?

****

the distribution function for the exponential distribution is $1-e^{-x/\theta}$

$$\therefore P(X \le 200) = 1 - e^{-\frac{200}{250}}$$

$$= 1 - e^{-.8} = 1 - .449 = .551$$

GAMMA

6-43

The failure time of a commercial electronic component has a gamma c.d.f. with parameters of 2 and 0.001/hour. A manufacturer claims to have a substitute component that has an exponential lifetime with a mean life of 2000 hours. The claim is based on the fact that both components have the same mean lifetime. For a reliability of 0.99, how long can each component be operated? Which component appears to be better ?

****

$Pr\{T > t_0\} = .99$, T = LIFETIME – FIND t_0

i) $T \sim \exp(2000)$; $f_e(t) = \frac{1}{2000} e^{-t/2000}$

$$Pr\{T > t_0\} = 1 - F_e(t_0) = e^{-t_0/2000} = .99$$

$$t_0 = -2000 \ln(.99) = 20.10$$

ii) $T \sim \text{GAMMA}(2, .001)$; $f_Y(t) = (.001)^2 t e^{-.001t}$

$$Pr\{T > t_0\} = \int_{t_0}^{\infty} f_Y(t)dt = \int_{t_0}^{\infty} (.001)^2 t e^{-.001t} dt$$

(LOOK THIS UP IN A HANDBOOK OR USE INTEGRATION BY PARTS)

$$= (.001 t_0 + 1) e^{-.001 t_0} = .99$$

$t_0 \simeq 148$ (TRIAL AND ERROR)

THE COMPONENT WITH THE GAMMA C.D.F. WOULD APPEAR TO BE BETTER.

6-44

The waiting time on line for a customer to be served at a bank is known to follow a Gamma distribution with mean of 5 minutes and standard deviation of 2.5 minutes. Find the probability of having to wait for more than 7 minutes. Leave your answer in terms of an integral.

****

Let r.v. X = Waiting time to be served

$$f_X(x) = \frac{1}{\beta^\alpha \Gamma(\alpha)} x^{\alpha-1} e^{-x/\beta} \qquad x > 0$$

$$\mu = \alpha\beta = 5 \qquad \frac{\sigma^2}{\mu} = \beta = \frac{6.25}{5} = 1.25$$

$$\sigma^2 = \alpha\beta^2 = (2.5)^2 = 6.25$$

$$\frac{\mu}{\beta} = \alpha = \frac{5}{1.25} = 4$$

$$Pr(X > 7) = \int_0^7 \frac{1}{(1.25)^4 \Gamma(4)} x^3 e^{-x/1.25} dx$$

$$= \frac{1}{(1.25)^4 3!} \int_0^7 x^3 e^{-x/1.25} dx$$

WEIBULL

6-45

Integrate the following:

$$\int_0^\infty x^3 \exp(-(x/2)^4)\ dx.$$

****

THIS INTEGRAL IS VERY CLOSE TO THE TOTAL INTEGRAL OF A WEIBULL P.D.F.

$$\int_0^\infty x^3 e^{-(x/2)^4} dx = 8\int_0^\infty (x/2)^3 e^{-(x/2)^4} dx = \frac{2\cdot 8}{4}\int_0^\infty \frac{4}{2}(x/2)^3 e^{-(x/2)^4} dx$$

$$= 4.$$

BETA

6-46

The PERT (Project Evaluation and Review Technique) project management method often assumes the Beta distribution to calculate project activity completion probabilities. If a Beta distribution with parameters of alpha = 4 and beta = 3 describes the fraction of maximum activity time required to actually perform the activity:

a) What is the mean fraction of maximum time required?

b) What is the probability that the activity will be finished by 80% of the maximum time?

****

a) $\mu = \alpha/(\alpha+\beta) = 0.571$

b) $P[x < 0.80] = \int_0^{0.8} \frac{\Gamma(7)}{\Gamma(4)\Gamma(3)} x^3(1-x)^2\,dx$

$= 60\int_0^{0.8} (x^3 - 2x^4 + x^5)\,dx = 0.900$

OTHER

6-47

Random variable X has the p.d.f. shown below. Find the value of "A" and the variance of X. Leave your answer in terms of integrals.

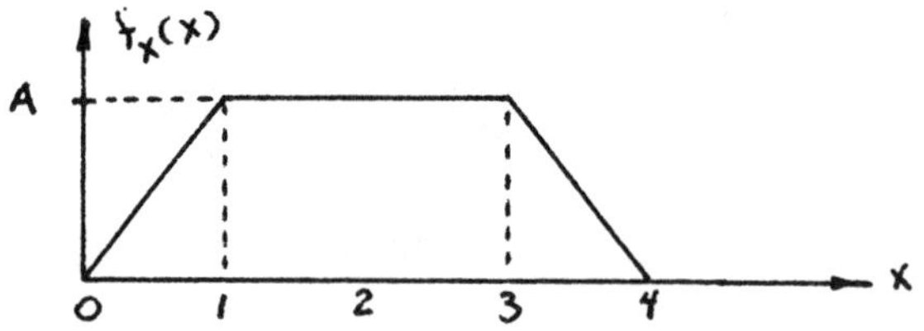

****

$$1 = \int_0^4 f_X(x)\,dx = \text{Area of Trapezoid} = \frac{1}{2}(2+4)A$$

$$A = \frac{1}{3}; \quad \mu_X = 2 \quad \text{By Symmetry of } f_X(x) \text{ about } x = 2$$

$$\sigma_X^2 = \int_0^4 (x-\mu_X)^2 f_X(x)\,dx$$

$$= \int_0^1 (x-2)^2 \frac{x}{3}\,dx + \int_1^3 (x-2)^2 \frac{1}{3}\,dx + \int_3^4 (x-2)^2 \frac{4-x}{3}\,dx$$

Since

$$f_X(x) = \begin{cases} x/3 & 0 \le x \le 1 \\ 1/3 & 1 \le x \le 3 \\ (4-x)/3 & 3 \le x \le 4 \end{cases}$$

6-48

The number of passengers arriving at a given bus stop is Poisson distributed with a mean rate of 1.2 passengers per minute.

Suppose that a bus (bus no. 2) arrives to pick up the waiting passengers 15 minutes after the previous bus (bus no. 1) has left this stop (assume that bus no. 1 had enough capacity to where no waiting passengers were left behind). The number of available seats on bus no. 2 is a random variable that depends on how many passengers got on board and got off at previous stops along the route. Assume that this random variable, denoted by X, is approximately normally distributed with a mean of 20 and a standard deviation of 5 seats. Let N be the number of passengers waiting at the stop at the time it is reached by bus no. 2.

1) What is the interpretation (in words) of the random variable (N-X)? Please limit your answer to one sentence.

2) Find the expected value of (N-X).

3) Assuming that N and X are independent, find the variance of (N-X) and of (3N-2X).

****

1) (N−X), if > 0, is the number of passengers that do not get a seat on the bus.

If (N−X) < 0, then its absolute value is the number of empty seats on the bus after it leaves the stop.

2) $E[N-X] = E[N] - E[X]$

N, the # of waiting passengers, is equal to the number of Poisson arrivals (mean rate = 1.2 pax/min) in a 15 minute interval; thus $E[N] = 1.2 \times 15 = 18$ passengers and $V[N] = 1.2 \times 15 = 18$

Thus $E[N-X] = 18 - 20 = \underline{\underline{-2}}$.

3) $V[N-X] = V[N] + V[X]$ (since X, N are independent)
$= 18 + (5)^2 = 18 + 25 = \underline{\underline{43}}$.

$V[3N-2X] = 9V[N] + 4V[X]$
$= 9 \times 18 + 4 \times 25 = 162 + 100 = \underline{\underline{262}}$.

6-49

A certain statistic follows the Chi-Square distribution with 20 degrees of freedom.

a) What numerical value will the sample statistic exceed with a probability of 0.9?

b) Find two values L and U such that the probability the statistic is less than L equals the probability it is greater than U and both of these probabilities are equal to 0.05.

****

Let r.v. X be distributed χ^2 with $\nu = 20$ d.f.

a)

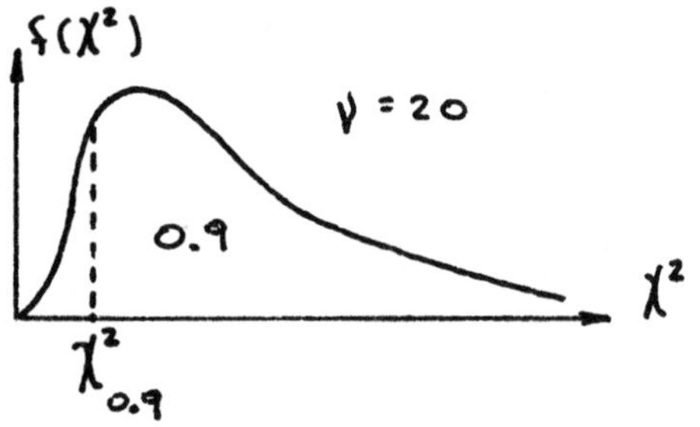

$$Pr(\chi^2 > \chi^2_{0.9}) = 0.9$$

$$\chi^2_{0.9} = 12.44 \text{ from Table}$$

b)

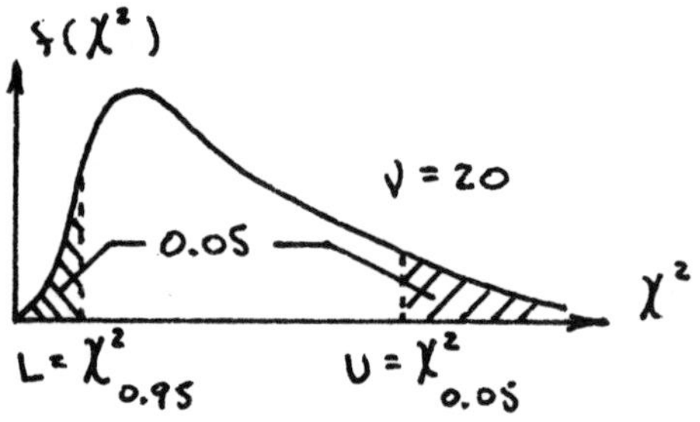

$$Pr(\chi^2 < L) = Pr(\chi^2 > U) = 0.05$$

$$L = \chi^2_{0.095} = 10.85$$

$$U = \chi^2_{0.05} = 31.41$$

7
MEANS

POINT ESTIMATES

7-1

An engineering society wishes to determine if the mean income of practicing electrical engineers exceeds that of civil engineers. Two independent random samples provided the following data:

	Electrical Engineers	Civil Engineers
sample size	76	58
mean	\$37,246	\$36,412
standard deviation	\$8,371	\$8,856

Letting $\bar{X}_1$ denote the mean income of a sample of 76 electrical engineers and $\bar{X}_2$ the mean income of a sample of 58 civil engineers, what is the distribution of $\bar{X}_1 - \bar{X}_2$? Make sure to state whether this distribution is exact or approximate. What are the best estimates of the moments (mean and variance) of this distribution?

****

By the Central Limit Theorem, $\bar{X}_1 \underset{\text{approx.}}{\sim} N\left(\mu_1, \frac{\sigma_1^2}{n_1}\right)$

$$\bar{X}_2 \underset{\text{approx.}}{\sim} N\left(\mu_2, \frac{\sigma_2^2}{n_2}\right)$$

where μ_i and σ_i^2 are the respective mean and variance of the population from which sample i was drawn and n_i is the sample size, $i = 1, 2$.

$$\bar{X}_1 - \bar{X}_2 \underset{\text{approx.}}{\sim} N\left[(\mu_1 - \mu_2), \left(\frac{\sigma_1^2}{n_1} + \frac{\sigma_2^2}{n_2}\right)\right]$$

Best estimates:

mean $$\hat{E}[\bar{X}_1 - \bar{X}_2] = \hat{\mu}_1 - \hat{\mu}_2 = 37{,}246 - 36{,}412 = \$834$$

variance $$\hat{V}[\bar{X}_1 - \bar{X}_2] = \frac{s_1^2}{n_1} + \frac{s_2^2}{n_2},$$

where s_1^2, s_2^2 are the respective sample variances,

$$\hat{V}[\bar{X}_1 - \bar{X}_2] = \frac{(8{,}371)^2}{76} + \frac{(8{,}856)^2}{58} = 2{,}274{,}241.2$$

std. dev. $$\sqrt{\hat{V}[\bar{X}_1 - \bar{X}_2]} = \$1508.1$$

7-2

If the size of a sample is 36 and the standard deviation of the mean is 2, what must the size of the sample become if the standard deviation of the mean is to be reduced to 1.2?

****

$$\sigma_{\bar{X}} = \sigma_X / \sqrt{n} \quad \text{or} \quad \sigma_X = \sqrt{n} \cdot \sigma_{\bar{X}}$$

Equating $(\sigma_X)_1 = (\sigma_X)_2$: $\sqrt{36} \cdot 2 = \sqrt{n_2}\,(1.2)$

$$\sqrt{n_2} = 10 \quad \text{or} \quad n_2 = 100$$

7-3

In 1650 the population of dodo birds on the island of Mauritius was 120 individuals. Ten years later it had dropped to 90. In 1690 only 36 birds were left. Calculate an average percent decrease in the population per decade to the nearest whole percent.

****

The geometric mean should be used to determine the least biased estimate of the percent decrease desired. This is accomplished by taking ratios of the population from decade to decade, as follows, realizing that no population figures are known for years 1670 and 1680 (call these "a" and "b"):

Year (Decade)	Population	Ratio
1650	120	
		90/120
1660	90	
		a/90
1670	a	
		b/a
1680	b	
		36/b
1690	36	

The geometric mean is then the fourth root of the product of the four ratios:

$$\text{Geometric mean} = \sqrt[4]{\frac{90}{120} \cdot \frac{a}{90} \cdot \frac{b}{a} \cdot \frac{36}{b}} \quad \text{or simply}$$

$$= \sqrt[4]{\frac{36}{120}}$$

$$= 0.74$$

In other words, the population in any decade is approximately 74% of what it was the previous decade, on average.

Percent decrease of the population per decade is thus

$$(1 - 0.74) \times 100\% = \underline{\underline{26\%}}$$

7-4

The sample mean

$$\bar{X} = \frac{1}{n} \sum_{i=1}^{n} X_i$$

of a set of n measurements $X_1, X_2, \ldots, X_n$ is taken as a point estimator for the mean value of the measurand. If each measurement is independent and has a standard deviation of σ, show that the standard deviation of $\bar{X}$ is $\sigma/\sqrt{n}$.

A gaging device produces a random error whose standard deviation is 1%. How many measurements should be averaged in order to reduce the standard deviation of the error to less than 0.05%?

****

$$\text{Variance}(\bar{x}) = \text{Var}\left(\tfrac{1}{n}(x_1 + x_2 + \cdots + x_n)\right)$$

$$= \frac{1}{n^2} \text{Var}(x_1 + x_2 + \cdots + x_n)$$

$$= \frac{1}{n^2}\left[\text{Var}(x_1) + \text{Var}(x_2) + \cdots + \text{Var}(x_n)\right]$$

(because x_i are independent measurements)

$$= \frac{n\sigma^2}{n^2} = \frac{\sigma^2}{n}$$

$\therefore$ std. deviation of $\bar{x}$ $= \sigma/\sqrt{n}$

With $\sigma = 1\%$ and $\frac{\sigma}{\sqrt{n}} < 0.05\%$ we have,

$$\frac{1}{\sqrt{n}} < 0.05 \Rightarrow \sqrt{n} > \frac{1}{0.05} = 20$$

Hence $n > 400$

More than 400 measurements have to be averaged.

INTERVAL ESTIMATES

7-5

The work experience of all employees at a large company is Normally distributed with mean μ = 12 years and standard deviation σ = 4.2 years. A committee of employees is to be selected to make recommendations to the president. To be completely fair, the committee will be randomly selected. How many members should there be on the committee to be 95% certain that the average work experience of the committee as a whole is greater than 10 years?

****

Let X_i = work experience of ith member

$\mu_{X_i} = 12$, $\sigma_{X_i} = 4.2$ $\quad i = 1, 2, \ldots, n$

where n is the number of members chosen

$$Pr(\bar{X} > 10) = 0.95$$

$$Pr\left[\frac{\bar{X} - \mu_{\bar{X}}}{\sigma_{\bar{X}}} > \frac{10 - \mu_{\bar{X}}}{\sigma_{\bar{X}}}\right] = 0.95 \quad \text{where } \mu_{\bar{X}} = \mu = 12$$

$$\sigma_{\bar{X}} = \frac{\sigma}{\sqrt{n}} = \frac{4.2}{\sqrt{n}}$$

$$Pr\left[Z > \frac{10 - 12}{4.2/\sqrt{n}}\right] = 0.95$$

But $Pr(Z > z_{0.95}) = 0.95$

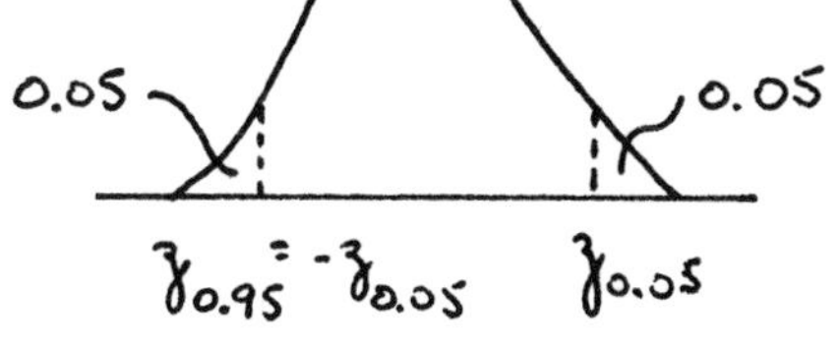

Therefore, $\dfrac{10-12}{4.2/\sqrt{n}} = z_{0.95} = -z_{0.05}$

$$-2\sqrt{n} = 4.2(-z_{0.05}) = 4.2(-1.645)$$

$n = 11.9$ $\qquad$ Ans. $n = 12$ people

7-6

A sample from an assumed normal distribution produced the values 9, 14, 10, 12, 7, 13, 12. a) What is the single best estimate of μ. b) What is the single best estimate of σ. c) Find an 80% confidence interval for μ.

****

a) $\mu \cong \overline{X} = 88/8 = 11$

b) $\sigma^2 \cong s^2 = \dfrac{n\sum x_i^2 - (\sum x_i)^2}{n(n-1)} = 5.143$

$\sigma \cong s = \sqrt{5.143} = 2.27$ mean added in

(c) FOR SMALL SAMPLES "t" distribution is used

80% CONF. LIMIT: $P(A < \mu < B) \geqq .80$

or $P\left(11 - \dfrac{t_{.10,7} \cdot s}{\sqrt{n}} < \mu < 11 + \dfrac{t_{.10,7} \cdot s}{\sqrt{n}}\right)$

$11 \pm 1.415\left(\dfrac{2.27}{\sqrt{8}}\right) = 11 \pm 1.13$ or

$P(9.87 < \mu < 12.13) \geq .80$

7-7

A research worker wishes to estimate the mean of a population by using a sample large enough that the probability will be 0.95 that the sample mean will not differ from the true mean by more than 25% of the standard deviation. How large a sample should be taken?

****

Pictorially:

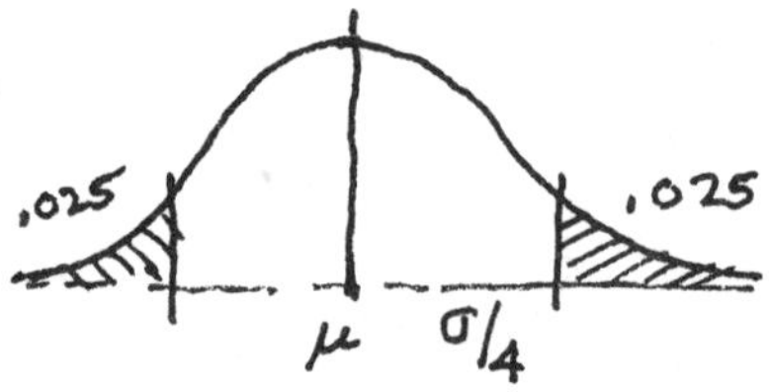

DIST. OF $\overline{X}$ has $\sigma_{\overline{x}} = \dfrac{\sigma}{\sqrt{n}}$

$\therefore Z_{.025} = 1.96 = \dfrac{\sigma/4}{\sigma/\sqrt{n}} = \dfrac{\sqrt{n}}{4}$

OR $n = (7.84)^2 \cong 62$

7-8

Sixteen randomly observed times (in minutes) for a certain task are as follows: 17.2, 18.7, 22.0, 19.3, 23.8, 20.1, 15.9, 24.2, 18.8, 19.0, 23.1, 22.0, 21.3, 18.4, 19.9, 20.3. Assuming that the population of such times is normally distributed, construct a 95% confidence interval for the population mean time to perform the task.

****

$n = 16$, σ unknown, population normal. therefore t distribution is appropriate.

$$\bar{x} = \frac{17.2 + 18.7 + \cdots + 20.3}{16} = \frac{324.0}{16} = 20.25$$

$s^2 = 5.488$, $s = 2.343$

95% C.I. for μ: $\left(\bar{x} \pm t_{\frac{\alpha}{2}, n-1} \frac{s}{\sqrt{n}}\right)$

$$= 20.25 \pm 2.131\left(\frac{2.343}{4}\right) = (19.00; 21.50)$$

7-9

Suppose that $\bar{x}$ and $\bar{y}$ are the means of two samples of size n each from a normal population with known variance σ^2. Determine n so that the probability will be about .99 that the two sample means will differ by less than σ.

****

A confidence interval for $(\mu_1 - \mu_2)$ is $Z_{.005} \cdot \sigma \sqrt{\frac{2}{n}}$ if $\sigma_1 = \sigma_2 = \sigma$. $\therefore Z_{.005} \cdot \sigma \cdot \sqrt{\frac{2}{n}} = \sigma$

$$2.576 \sqrt{\frac{2}{n}} = 1$$

Solving for n: $n = 2(2.576)^2 = 13.3$

OR ROUNDING OFF $n = 14$

7-10

The salaries of executives of a major corporation is under investigation. Ten executives were selected at random and their salaries are given below.

$45,000 $52,000 $61,000 $73,000 $55,000 $47,000 $84,000
$69,000 $51,000 $60,000

Assuming the salaries of all executives are Normally distributed, find a
a) 95% 2-sided confidence interval
b) 99% 2-sided confidence interval
for the mean salary of all executives.

****

The $100(1-\alpha)\%$ 2-sided confidence interval for μ, the unknown mean salary of all executives, is

$$\bar{x} \pm t_{\alpha/2} \frac{s}{\sqrt{n}}$$

The t-distribution is used since the population std. deviation σ is unknown.

Based on the data, $\bar{x} = \dfrac{\Sigma x_i}{n} = 59{,}700.$

$$s = \sqrt{\frac{\Sigma (x_i - \bar{x})^2}{n-1}} = 12{,}428.$$

a) $\alpha = 0.05$, $t_{\alpha/2} = t_{0.025} = 2.262$ with $\nu = 10-1 = 9$ d.f.

$$95\%\text{ C.I.} = 59{,}700 \pm 2.262 \frac{(12{,}428)}{\sqrt{10}}$$

$$= 59{,}700. \pm 8{,}890.$$

b) $\alpha = 0.01$, $t_{\alpha/2} = t_{0.005} = 3.250$ with $\nu = 9$ d.f.

$$99\%\text{ C.I.} = 59{,}700. \pm 3.250 \frac{(12{,}428)}{\sqrt{10}}$$

$$= 59{,}700. \pm 12{,}772.$$

7-11

The weights of full boxes of a certain kind of cereal are normally distributed with a standard deviation of 0.27 ounce. If a sample of 15 ramdomly selected boxes produced a mean weight of 9.87 ounces, find

a) the 95% confidence interval for the true mean weight of a box of this cereal.

b) the 99% confidence interval for the true mean weight of a box of this cereal.

c) what effect does the increase in the level of confidence have on the width of the confidence interval?

****

$\sigma = 0.27$ oz

$n = 15$

$\bar{X} = 9.87$ oz

a) 95% CONFIDENCE IMPLIES A TWO SIDED INTERVAL WITH $\alpha/2 = 0.05/2 = 0.025$ IN EACH TAIL OF THE P.D.F.

$$Pr\{-z_{.025} \le Z \le z_{.025}\} = .95$$

$$\pm z_{.025} = \pm 1.96$$

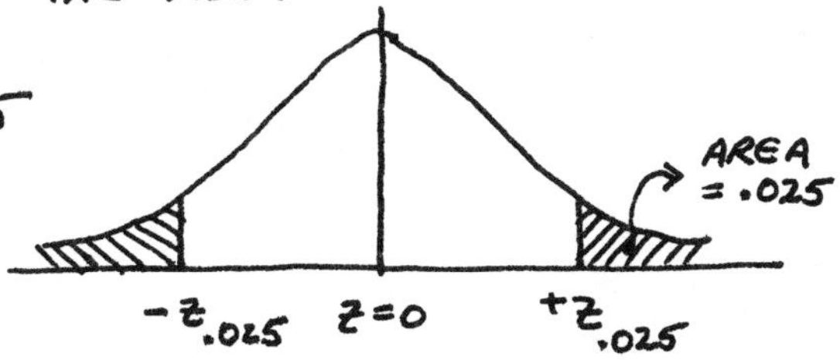

$$Z = \frac{\bar{X}-\mu}{\sigma/\sqrt{n}}, \quad \bar{X} - z_{.025}\,\sigma/\sqrt{n} \le \mu \le \bar{X} + z_{.025}\,\sigma/\sqrt{n}$$

$$9.87 - .14 \le \mu \le 9.87 + .14$$

$$9.73 \le \mu \le 10.01$$

b) THE PROBLEM IS THE SAME EXCEPT $\alpha = 0.01$, $\alpha/2 = .005$

$$\pm z_{.005} = \pm 2.575; \quad 9.87 - .18 \le \mu \le 9.87 + .18$$

$$9.69 \le \mu \le 10.05$$

c) IT WIDENS THE LENGTH OF THE CONFIDENCE INTERVAL.

7-12

The angular resolution (smallest change in angle that could be read) of a resolver (a device that measures angular motions) was tested sixteen times independently and recorded in degrees as follows:

0.11, 0.12, 0.09, 0.10, 0.10, 0.14, 0.08, 0.08,
0.13, 0.10, 0.10, 0.12, 0.08, 0.09, 0.11, 0.15

If the standard deviation of the angular resolution of this type of resolvers is known to be $\sigma = 0.01°$, what are the odds that the mean value μ of the resolution of this type of resolvers would lie within 5% of the sample? Assume that the angular resoltuion is normally distributed.

****

Consider the random variable $\bar{X}$ which is the average of the n measurements $X_1, X_2, \cdots, X_n$;

$$\bar{X} = \frac{1}{n} \sum_{i=1}^{n} X_i$$

Since X_i are normal with mean μ ad std. deviation σ, $\bar{X}$ would also be normal with mean μ and std. deviation $\sigma/\sqrt{n}$. Hence the random variable

$$Z = \frac{\bar{X} - \mu}{\sigma/\sqrt{n}}$$

would be normally distributed with zero mean and unity std. deviation.

Suppose

$$P(-z_0 < Z < z_0) = p$$

for positive z_0. Note that if z_0 is known p could be read from standard Gaussian tables. Substitute for Z;

$$P\left(-z_0 < \frac{\bar{X} - \mu}{\sigma/\sqrt{n}} < z_0\right) = p$$

$$= P\left(\bar{X} - \frac{z_0 \sigma}{\sqrt{n}} < \mu < \bar{X} + \frac{z_0 \sigma}{\sqrt{n}}\right)$$

It should be kept in mind that μ is a deterministic yet unknown parameter and $\bar{X}$ is the random variable. The data gives just a sample value for $\bar{X}$. This sample value is calculated to be

$$\bar{X} = \frac{1}{16}(0.11 + 0.12 + \cdots + 0.11 + 0.15) = 0.10625$$

We need $$\frac{z_0 \sigma}{\sqrt{5}} = 5\% \text{ of } \bar{X} = \frac{5}{100} \times 0.10625$$

Now with $\sigma = 0.01$ we have

$$z_0 = \frac{5}{100} \times 0.10625 \times \frac{\sqrt{5}}{0.01} = 2.125$$

From tables

$$P(-2.125 < z < 2.125)$$

$$= 1 - 2 \times (1 - 0.98332) = \underline{\underline{0.9664}}$$

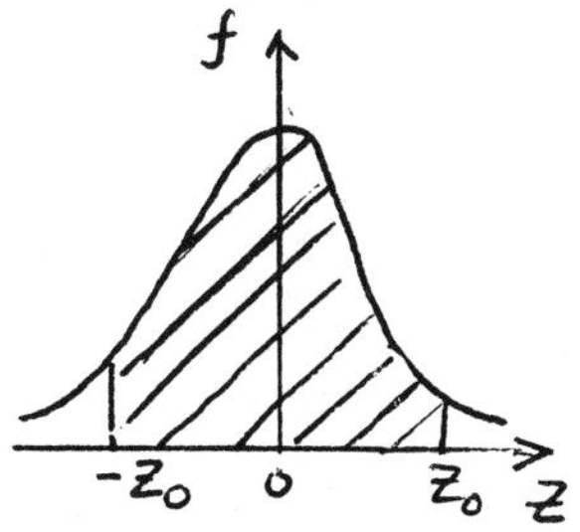

The odds are about 96% that the true mean would lie within 5% of the sample mean.

7-13

In a study of material costs, a random sample of 50 cases of a particular model had a mean of \$973.25 and a standard deviation of \$72.49.

a) What is the probability that an error of no more than \$10 is made when estimating the true average material cost of this model as \$973.25?

b) How many additional cases should you sample to be able to assert with a probability of 0.95 that the sample mean will be within \$10 of the true mean?

****

ASSUMPTION: COSTS ARE NORMALLY DISTRIBUTED

$\bar{X} = 973.25 \qquad s = 72.49 \qquad n = 50$

a) $\mu = \bar{X} \pm z_{\alpha/2} \dfrac{s}{\sqrt{n}}$ CONFIDENCE INTERVAL ASSUMING $s \approx \sigma$ SINCE $n > 30$

$$|\mu - \bar{x}| = E = z_{\alpha/2} \frac{s}{\sqrt{n}}$$

$$z_{\alpha/2} = \frac{E\sqrt{n}}{s} = \frac{10\sqrt{50}}{72.49} = 0.9755$$

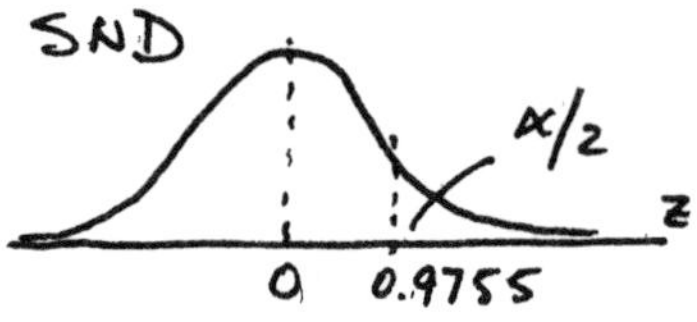

$Pr(z < 0.9755) = 0.8353$

$Pr(z > 0.9755) = 0.1647 = \alpha/2$

$\therefore \quad \alpha = 0.3294$

$$1 - \alpha = 0.6706 = P(|E| < \$10)$$

b) $n \geq \left(z_{\alpha/2} \cdot \dfrac{s}{E}\right)^2 = \left(1.96 \dfrac{72.49}{10}\right)^2 = 201.9$

$n = 202$

MUST SAMPLE $202 - 50 = 152$ ADDITIONAL CASES

7-14

The mean of the grades of 40 freshmen is used to estimate the true average grade of the freshman class. If the standard deviation of all grades is known to be 10, find the probability that the sampling error is more than 2.5.

****

Let X_i = Grade of ith student in sample, $i = 1, 2, 3, \ldots, 40$

μ = True (unknown) average grade of entire class

The sampling statistic is $\bar{X}$ and the sampling error is given by $|\bar{X} - \mu|$

$$\Pr(|\bar{X}-\mu| > 2.5) = 1 - \Pr(|\bar{X}-\mu| \le 2.5)$$

$$= 1 - \Pr(-2.5 \le \bar{X}-\mu \le 2.5)$$

$$= 1 - \Pr\left[\frac{-2.5}{\sigma_{\bar{X}}} \le \frac{\bar{X}-\mu}{\sigma_{\bar{X}}} \le \frac{2.5}{\sigma_{\bar{X}}}\right]$$

But $\mu_{\bar{X}} = \mu$, therefore $\dfrac{\bar{X}-\mu}{\sigma_{\bar{X}}} = \dfrac{\bar{X}-\mu_{\bar{X}}}{\sigma_{\bar{X}}} = Z \sim N(0,1)$ by C.L.T.

and $\sigma_{\bar{X}} = \dfrac{\sigma}{\sqrt{n}} = \dfrac{10}{\sqrt{40}}$

$= \sqrt{5/2}$

$$= 1 - \Pr\left[\frac{-2.5}{\sqrt{5/2}} \le Z \le \frac{2.5}{\sqrt{5/2}}\right]$$

$$= 1 - \left\{F_Z\left(\frac{2.5}{\sqrt{5/2}}\right) - F_Z\left(\frac{-2.5}{\sqrt{5/2}}\right)\right\}$$

$$= 1 - \{F_Z(\sqrt{2.5}) - F_Z(-\sqrt{2.5})\}$$

$$= 1 - \{F_Z(\sqrt{2.5}) - [1 - F_Z(\sqrt{2.5})]\}$$

$$= 2\{1 - F_Z(\sqrt{2.5})\}$$

$$= 0.114$$

7-15

From a large group of people, 4 persons were weighed as: 137, 147, 158, and 170 lbs. Also, from another large group of people, 5 persons were weighed as: 132, 145, 162, 166, and 175 lbs. Determine a 95% confidence interval for the difference between the two class means, $\Delta\mu = \mu_1 - \mu_2$.

****

This is the case where population variances are unknown. Let X_1 represent 5-person sample and X_2 4-person sample. Tabulate computations as follows.

Group I			Group II		
X_1	$(X_1-\bar{X}_1)$	$(X_1-\bar{X}_1)^2$	X_2	$(X_2-\bar{X}_2)$	$(X_2-\bar{X}_2)^2$
132	−24	576	137	−16	256
145	−11	121	147	−6	36
162	6	36	158	5	25
166	10	100	170	17	289
175	19	361			

$$\bar{X}_1 = \frac{1}{n_1}\sum X_1 = \frac{780}{5} = 156$$

$$\bar{X}_2 = \frac{1}{n_2}\sum X_2 = \frac{612}{4} = 153$$

$$\sum(X_1-\bar{X}_1)^2 = 1194; \quad \sum(X_2-\bar{X}_2)^2 = 606$$

The pooled variance is given as:

$$S_p^2 = \frac{1}{(n_1 + n_2 - 2)} \left[\sum^{n_1} (X_1 - \bar{X}_1)^2 + \sum^{n_2} (X_2 - \bar{X}_2)^2 \right]$$

$$S_p^2 = \frac{1}{7}(1194 + 606) = 257$$

With 7 degrees of freedom, ($n_1 + n_2 - 2 = 5+4-2 = 7$), from Student's t distribution tables one has: $t_{0.025} = 2.365$.

The 95% confidence interval for the difference of two means is given as:

$$\Delta\mu = \mu_1 - \mu_2 = (\bar{X}_1 - \bar{X}_2) \pm t_{0.025} S_p \sqrt{\frac{1}{n_1} + \frac{1}{n_2}}$$

$$= 156 - 153 \pm 2.365 \sqrt{257} \sqrt{\frac{1}{5} + \frac{1}{4}}$$

$$= 3 \pm 25.4$$

Or, $-22.4 < \Delta\mu < 28.4$

7-16

A sample drawn from a Normal population resulted in the following values:
$X_1 = 7$, $X_2 = 10$, $X_3 = 12$, $X_4 = 9$, $X_5 = 10$
Find the population standard deviation if the 90% confidence interval for the unknown mean is (8,11.2).

****

$$\sum_{i=1}^{5} X_i = 7 + 10 + 12 + 9 + 10 = 48 \qquad \bar{X} = \frac{48}{5} = 9.6$$

$$90\% \text{ C.I. for } \mu = \bar{X} \pm z_{.05} \frac{\sigma}{\sqrt{5}} = 9.6 \pm 1.645 \frac{\sigma}{\sqrt{5}}$$

From the problem statement, the 90% C.I. is between 8 and 11.2, i.e.

$$90\% \text{ C.I. for } \mu = \frac{8 + 11.2}{2} \pm \frac{11.2 - 8}{2} = 9.6 \pm 1.6$$

Equating the above intervals,

$$9.6 \pm 1.645 \frac{\sigma}{\sqrt{5}} = 9.6 \pm 1.6 \implies \sigma = \frac{1.6\sqrt{5}}{1.645} = 2.17$$

7-17

A distribution {x} is normal with mean 20, standard deviation 5. If samples of 8 of the x's are randomly taken and their average calculated, what are the values U and L over which 85% of the $\bar{x}$ will range, with equal probability of $\bar{x}$ being above U or below L.

****

VALUE OF Z FOR WHICH TAIL AREA IS $\frac{1-.85}{2} = Z_{.075} = 1.44$

OR THE CONFIDENCE LIMITS ARE $20 \pm 1.44\left(\frac{5}{\sqrt{8}}\right)$

$$= 20 \pm 2.55$$

$$\therefore L = 20 - 2.55 = 17.45 \; ; \; U = 20 + 2.55 = 22.55$$

7-18

A marketing firm wishes to estimate the average living area of homes in a residential area. Based on a survey in a similar neighborhood, the living area can be assumed Normally distributed with a sample standard deviation of 175 square feet. The firm would like to be 99% sure their estimate of the average living area is off by no more than 100 square feet. How large a sample size is required?

****

Let μ = Average (unknown) living area

$E_{MAX} = t_{\alpha/2} \dfrac{s}{\sqrt{n}}$ where E_{MAX} is the maximum error in using $\bar{x}$ to estimate μ

$\alpha = 0.01$ $t_{0.005}$ cannot be determined since we don't know $\nu = n-1$ degrees of freedom

Solving for n, $n = \left(\dfrac{t_{\alpha/2}\, s}{E_{MAX}}\right)^2 = \left[\dfrac{t_{\alpha/2}\,(175)}{100}\right]^2$

n is obtained by trial & error,

$n = 25$: $25 = [t_{0.005}(1.75)]^2$ $t_{0.005} = 2.797$ with $\nu = 24$ d.f.

$25 = [2.797(1.75)]^2 = 23.96$

$n = 20$: $20 = [2.861(1.75)]^2$ $t_{0.005} = 2.861$ with $\nu = 19$ d.f.

$= 24.2$

$n = 24$: $24 = [2.807(1.75)]^2$ $t_{0.005} = 2.807$ with $\nu = 23$ d.f.

Ans.

$= 24.1$

Note, the t-distribution was used since the population std. deviation is unknown.

7-19

The following sample data were obtained for the time (in minutes) required to complete a particular task on a construction site:

11.2	18.5	8.7	12.4	13.5
9.9	12.9	15.4	12.6	16.7
10.2	10.5	14.4	17.7	15.5

a) Construct a 95% confidence interval estimate of the mean task completion time. You may assume that completion times are normally distributed.

b) Without performing any additional calculations, would a 99% C.I. for the mean be shorter or longer than the one obtained in part (a)? Justify your answer (in words).

c) Assuming that the population standard deviation was actually known and equal to the calculated sample standard deviation for the above data, how would the length of the interval obtained in part (a) be affected? Do not perform any additional calculations, and justify your answer (in words).

d) Under the assumptions of part (c), what is the minimum sample size needed if the length of the 95% interval is to be 1.8 minutes.

****

a) Let X be the task completion time. $X \sim N(\eta_x, \sigma_x^2)$

X is normal, sample size $n = 15$ and pop. variance σ_x^2 unknown.

Let $\bar{X}$ and s denote the sample mean and sample standard deviation, respectively. For the given data, $\bar{X} = 13.34$ and $s = 2.98$.
C.I. constructed based on the statistic

$$\frac{\bar{X} - \eta_x}{s/\sqrt{n}} \sim t(n-1) \quad [\text{t-distributed}, n-1 \text{ d.f.}]$$

yielding the 95% C.I. for η_x as follows:

$$\left[\bar{X} - t_{0.025,14}\frac{s}{\sqrt{n}}, \ \bar{X} + t_{0.025,14}\frac{s}{\sqrt{n}}\right]$$

$$\left[13.34 - 2.145 \cdot \frac{2.98}{\sqrt{15}}, \ 13.34 + 2.145 \cdot \frac{2.98}{\sqrt{15}}\right]$$

95% C.I. → [11.69, 14.99].

b) The 99% C.I. would be longer

because of the higher confidence level.

c) If the population std. dev. σ_x was known,

the appropriate statistic and distribution to construct

the confidence interval would be $\dfrac{\bar{X} - \mu_x}{\sigma_x/\sqrt{n}} \sim N(0, 1)$.

Since the std. normal dist. is "tighter" than the t-distribution

(smaller variance), the resulting interval will be shorter

than the one obtained in part (a).

d) interval length $= 2\, z_{0.025} \cdot \dfrac{\sigma_x}{\sqrt{n}} = 1.8$

$$\Rightarrow \sqrt{n} = \frac{2\, z_{0.025} \cdot \sigma_x}{1.8}$$

$$= \frac{2 \times 1.96 \times 2.98}{1.8} = 42.12$$

minimum sample size is 43 observations.

7-20

From a large group of people, 4 persons were weighed as: 139, 152, 160, and 173 lbs. Calculate a 95% confidence interval for the whole group mean, μ.

****

This is a case of t-distribution when σ^2 is unknown. Here, we have sample size $n = 4$, and degrees of freedom, d.f. = $n-1 = 4-1 = 3$. From the Student's t distribution tables, one has

$$t_{0.025} = 3.182$$

Let X represent the weight of a person, then necessary computation is tabulated as follows.

X	$X-\bar{X}$	$(X-\bar{X})^2$
139	−17	289
152	−4	16
160	4	16
173	17	289

$$\bar{X} = \frac{1}{n}\Sigma X = \frac{624}{4}$$

$$\bar{X} = 156$$

$$S^2 = \frac{1}{n-1}\Sigma(X-\bar{X})^2$$

$$S^2 = \frac{1}{3}(610) = 203.3$$

The 95% confidence interval is given as:

$$\mu = \bar{X} \pm t_{0.025} \frac{S}{\sqrt{n}}$$

$$\mu = 156 \pm 3.182 \times \frac{\sqrt{203.3}}{\sqrt{4}}$$

or, $\mu = 156 \pm 22.7$

or, $133.3 < \mu < 178.7$

HYPOTHESES TESTS: ONE MEAN

7-21

An aircraft parts manufacturer claims that the average weight of one of its parts is no more than 15 pounds, with a standard deviation of one pound. A particular customer accepts the standard deviation claim but decides to randomly select 49 of the parts to test the mean claim, setting the probability of a type I error at 1%. The customer accepts the burden of proof for the experiment. Determine the probability that the customer will accept the manufacturer's claim for mean weight if the manufacturer is actually producing parts with an average weight of 15.3 pounds.

****

$H_0: \mu = 15$ Large sample, Z appropriate

$H_1: \mu > 15$ Reject H_0 if $\bar{X} > C$ (criterion)

where $C = \mu_0 + Z_\alpha \frac{\sigma}{\sqrt{n}} = 15 + (2.33)\left(\frac{1}{7}\right) = 15.33$

$$\beta = P(\text{Type II error}) = P(\bar{X} < 15.33) \text{ when } \mu = 15.30$$

$$= P\left(z < \frac{15.33 - 15.30}{1/7}\right) = P(z \le 0.21) = 0.5832$$

7-22

A VENDOR CLAIMS THAT HERE ARE 64 OUNCES OF SOAP IN HIS "GIANT ECONOMY" SIZE BOX. PAST EXPERIENCE HAS SHOWN THAT THE VARIANCE ON FILLING BOXES TO BE 0.09 OZ^2. A SAMPLE OF 6 BOXES ARE WEIGHED AND THE AVERAGE AMOUNT OF SOAP WAS FOUND TO BE 63.25 OUNCES.

SHOULD WE ACCEPT THE VENDOR'S CLAIM ?

****

$\alpha = 0.05$ $\quad \sigma^2 = 0.09\ oz^2$ $\quad \sigma = 0.3\ oz$ $\quad N = 6$

$\sigma_{\bar{x}} = \frac{0.3}{\sqrt{6}} = 0.122$ $\quad \bar{x} = 63.25\ oz$

$H_0 : \mu = 64$ $\quad H_1 : \mu < 64$

$\eta = 6 - 1 = 5$ $\quad t_{5,\,0.05} = 2.015$

CRITICAL VALUE $C = 64 - t_{\eta,\alpha}\, \sigma_{\bar{x}}$

$C = 64 - 2.015\,(.122) = 63.75$

REJECT THE CLAIM

7-23

We wish to test the hypothesis that the mean IQ of the students in a school system is 100. Using $\sigma = 15$, $\alpha = .05$ and a sample of 25 students in a two sided test, the sample value $\bar{x}$ is computed. Find: a) the range of sample values of $\bar{x}$ for which we would accept the hypothesis, b) If the true IQ of the children is 105, find the probability of falsely accepting H_o: IQ = 100.

****

$H_0: \mu = 100$ with $H_1: \mu \neq 100$, $\alpha = .05$, $n = 25$

a) $100 \pm 1.96 \left(\frac{15}{\sqrt{25}}\right) = 100 \pm 5.88$ < 105.88, 94.12

b)

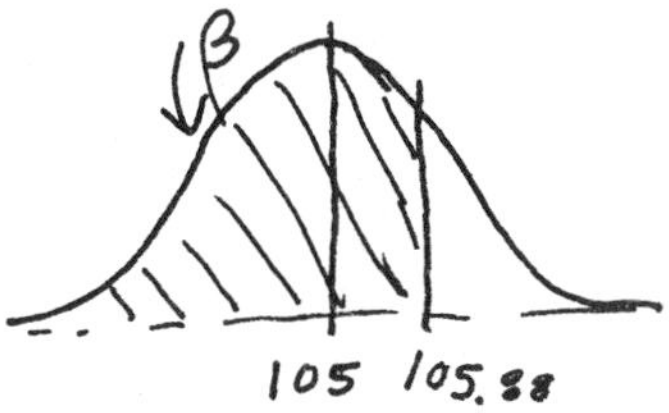

$$Z = \frac{105.88 - 105}{15/\sqrt{25}} = .29$$

$$\beta = .6141$$

7-24

Based upon a random sample of size 100 with an average of 3.4 minutes and a standard deviation of 2.8 minutes, is A.T.&T.'s claim that the average telephone call is 4 minutes true with a confidence of 95%?

****

$n = 100$
$\bar{X} = 3.4$ MIN
$S = 2.8$ MIN
$\alpha = 0.05$

$H_0: \mu = 4$ MIN
$H_A: \mu \neq 4$ MIN

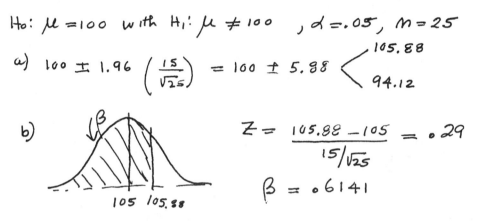

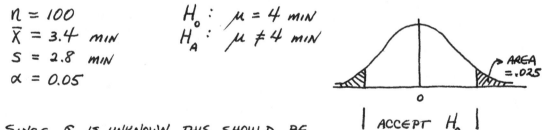

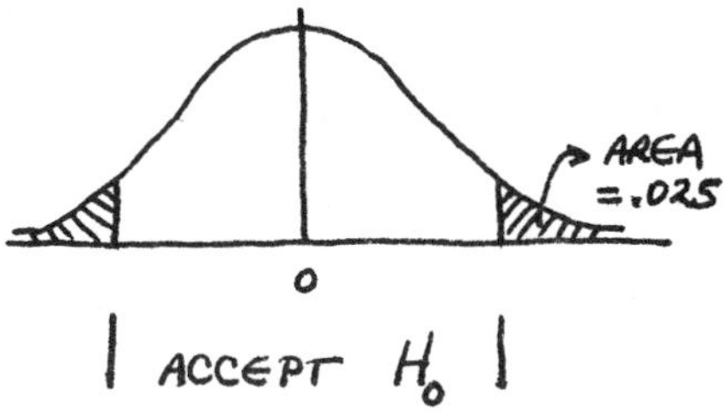

SINCE σ IS UNKNOWN THIS SHOULD BE A t-DISTRIBUTION; HOWEVER, SINCE $n = 100$ THE NORMAL C.D.F. IS USED.

$$z = \frac{\bar{X} - \mu}{s/\sqrt{n}} = \frac{3.4 - 4}{2.8/10} = -2.14$$

$$\pm z_{\alpha/2} = \pm z_{.025} = \pm 1.96$$

SINCE $-2.14 < -1.96$, REJECT H_0.

7-25

A manufacturer of athletic footwear claims that the mean life of his product will exceed 50 hours. A random sample will be tested and the claim accepted if the sample mean exceeds 54 hours. Assuming there is a 5% risk of false acceptance, find the required sample size. The shoe life is Normally distributed with standard deviation of 5 hours. Use $H_o: \mu=50$ and $H_1: \mu>50$.

****

$H_0: \mu = 50$ r.v. X = Shoe life, $X \sim N(\mu, \sigma)$

$H_1: \mu > 50$ $\sigma = 5$

Under the null hypothesis, H_0

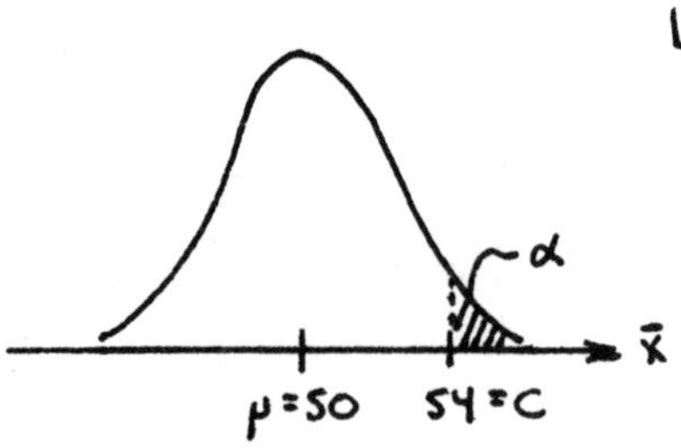

$$\Pr(\bar{x} > C \mid \mu = 50) = \alpha = 0.05$$

where $C = 54$, i.e. critical value

$$\Pr\left[\frac{\bar{x}-\mu}{\sigma/\sqrt{n}} > \frac{C-50}{5/\sqrt{n}}\right] = 0.05$$

$$\Pr\left[z > \frac{54-50}{5/\sqrt{n}}\right] = 0.05$$

But $\Pr(z > z_\alpha) = \alpha \implies \dfrac{54-50}{5/\sqrt{n}} = z_{0.05} = 1.645$

Solving for n yields $n = 4.3$

Ans. Sample Size = 5

Note, the sample size is relatively small because the critical value is significantly larger than the test value, 50. If the critical value were changed to 51 instead of 54, the required sample size would be 68.

7-26

THE THRUST TO SAFELY LIFT A SPACE CRAFT OFF THE LAUNCH PAD IS 4,000,000 LBS WITH A STANDARD DEVIATION OF 200,000 LBS. A COMPANY HAS COME UP WITH A NEW TYPE OF ROCKET WHICH IS CHEAPER THAN THE CURRENT MODEL. THEY CLAIM IT IS "JUST AS GOOD AS A ...."

A) STATE THE NULL AND ALTERNATIVE HYPOTHESIS IN WORDS.

B) SET UP THE TEST OF HYPOTHESIS TABLE, INDICATING THE TWO TYPES OF ERRORS

C) STATE THESE TWO TYPES OF ERRORS IN RELATION TO THIS SPECIFIC PROBLEM

D) WHICH TYPE OF ERROR IS MORE SERIOUS ?

E) IF 8 ENGINES WERE TESTED AND THE AVERAGE THRUST WAS 3,875,000 WHAT DECISION DO WE MAKE AT AN ALPHA = 0.05 ?

****

H_0: THE NEW ROCKET HAS AS MUCH THRUST AS THE OLD.

H_1: THE NEW ROCKET HAS *LESS* THRUST AS THE OLD.

$\mu_0 = 4 \times 10^6$ lbs

$\sigma^2 = .2 \times 10^6$ lbs^2

$\bar{X} = 3.875 \times 10^6$ lbs

$N = 8$

$H_0: \mu_1 = \mu_0 \quad H_1: \mu_1 < \mu_0$

$\mu_1 = 4 \times 10^6 \quad \mu_1 < 4 \times 10^6$

	H_0	H_1
ACCEPT H_0	—	II
ACCEPT H_1	I	—

TYPE I ERROR : WE SAY THAT THE THRUST IS LESS THAN 4 MILLION LBS WHEN IT ACTUALLY IS.

TYPE II ERROR : WE SAY THAT THE THRUST IS 4 MILLION LBS WHEN IT IS ACTUALLY LESS.

THE TYPE II ERROR IS MORE SERIOUS

$$t = \frac{3.875 - 4}{.2 \sqrt{8}} = -1.768$$

$$-t_{7,.05} = -1.895 \longrightarrow \text{ACCEPT } H_0$$

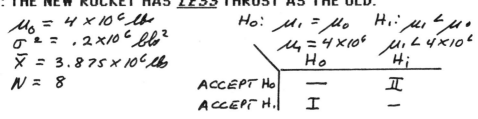

7-27

You want to test the hypothesis that the weekly grocery bill for a couple without children is \$57 against the alternative that it is \$61. From previous experience it is known that the distribution is normal with a standard deviation, σ = \$15.

a) If the probability of a type I error is to be 5% and the probability of a type II error is to be 10%, find the required sample size and the corresponding critical value.

b) If you are limited to a sample of size 25, determine the probability of a type I error if the critical value is set at \$59.50.

****

$H_0: \mu = 57 \qquad H_1: \mu = 61 \qquad \sigma = 15$

a) $\alpha = 0.05$, THEN $z_\alpha = 1.645$

$\beta = 0.10$, THEN $z_\beta = 1.28$

$$n \geq \left[\frac{z_\alpha \sigma_0 + z_\beta \sigma_1}{\mu_1 - \mu_0}\right]^2 = \left[\frac{(1.645)(15) + (1.28)(15)}{61 - 57}\right]^2$$

$n \geq 120.3$, THEN $n = 121$

$$X_c = \mu_0 + z_\alpha \frac{\sigma}{\sqrt{n}} = 57 + (1.645)\frac{(15)}{\sqrt{121}}$$

$$= \underline{59.24}$$

b) α = P(TYPE I ERROR) = P(REJECT H_0 / TRUE)

$= P(\bar{X} > 59.5 \mid \mu = 57, \sigma = 15, n = 25)$

$$= P\left(z > \frac{59.5 - 57}{15/\sqrt{25}}\right) = P(z > 0.833)$$

$= 1 - P(z < 0.833) = 1 - 0.7958$

$= 0.2042$

7-28

Engineers at the State Highway Department consider a highway to have reached an unacceptable congestion level when the mean number of vehicles per hour per lane during the peak period (also known as mean peak hourly traffic volume) is equal to or greater than 1800 vehicles/hr. A sample of traffic counts taken during the peak period over 20 days resulted in a sample mean of 1780 vehicles/hr and sample standard deviation of 150 vehicles/hr. It can be assumed that peak hourly traffic volume is normally distributed.

a) A highway will be considered unacceptable unless there is sufficient evidence to the contrary and the probability of a type I error does not exceed 0.05. After specifying the appropriate hypotheses, set up the critical region for this test. Please define clearly all symbols used.

b) Based on the above sample information, what can you conclude regarding the acceptability of the congestion levels for the highway under consideration?

****

a) Let X denote the peak hourly traffic volume (on a given day).

$X \sim N(\eta_x, \sigma_x^2)$; both η_x (pop. mean) and σ_x^2 (pop. variance) are unknown.

$H_0: \eta_x \geq 1800$ vs. $H_1: \eta_x < 1800$; $\alpha = 0.05$

Since pop. is normal, pop. variance is unknown, and sample size n is small, a t-test should be used.

test statistic: $\dfrac{\bar{X} - \eta_x}{S/\sqrt{n}} \sim t(n-1)$, where $\bar{X}$ = sample mean, S = sample std. dev.

critical region: reject H_0 if $\dfrac{\bar{X} - \eta_x}{S/\sqrt{n}} < -t_{0.05,19}$,

i.e., if $\bar{X} < 1800 - 1.729 \times \dfrac{150}{\sqrt{20}}$

Thus, if $\bar{X} < 1742$, reject H_0.

b) For given sample, $\bar{X} = 1780 > 1742 \Rightarrow$ DO NOT REJECT H_0;

Highway congestion is not acceptable on the facility under consideration.

7-29

The following test has been established.

$H_o: \mu = 100$

$H_1: \mu \neq 100$

The population standard deviation $\sigma = 15$. The probabilities of Type I and Type II errors are

Pr(Type I error) = α = 0.01

Pr(Type II error) = Pr(Accept $H_o | \mu = 110$) = β = 0.15

Find the critical value and sample size. State the decision rule.

****

Under the Null hypothesis, i.e. H_o is true

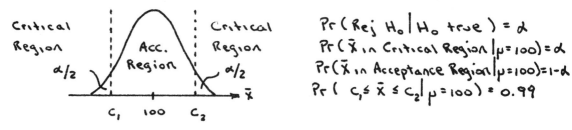

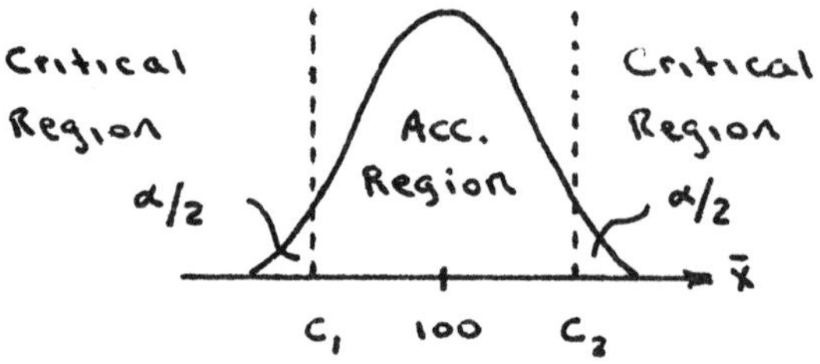

$Pr(\text{Rej } H_o | H_o \text{ true}) = \alpha$

$Pr(\bar{X} \text{ in Critical Region} | \mu = 100) = \alpha$

$Pr(\bar{X} \text{ in Acceptance Region} | \mu = 100) = 1 - \alpha$

$Pr(c_1 \le \bar{X} \le c_2 | \mu = 100) = 0.99$

Assuming the statistic $\bar{X}$ is Normally distributed with $\mu_{\bar{X}} = \mu_o = 100$, $\sigma_{\bar{X}} = \dfrac{\sigma}{\sqrt{n}} = \dfrac{15}{\sqrt{n}}$

$$Pr\left[\frac{c_1 - 100}{15/\sqrt{n}} \le \frac{\bar{X} - \mu_{\bar{X}}}{\sigma_{\bar{X}}} \le \frac{c_2 - 100}{15/\sqrt{n}}\right] = 0.99$$

$$Pr\left[\frac{(c_1 - 100)\sqrt{n}}{15} \le Z \le \frac{(c_2 - 100)\sqrt{n}}{15}\right] = 0.99$$

But $Pr(-z_{\alpha/2} \le Z \le z_{\alpha/2}) = 1 - \alpha$

Therefore, $$\frac{(c_2 - 100)\sqrt{n}}{15} = z_{\alpha/2} = z_{0.005} = 2.58$$

Under the alternate hypothesis, i.e. H_1 is true

$Pr(\text{Acc } H_o | H_o \text{ false}) = Pr(\bar{X} \text{ in Acc Region} | H_1 \text{ true}) = \beta$

$= Pr(c_1 \le \bar{X} \le c_2 | \mu = \mu_1 = 110) = 0.15$

$$Pr\left[\frac{c_1 - 110}{15/\sqrt{n}} \le Z \le \frac{c_2 - 110}{15/\sqrt{n}}\right] = 0.15$$

The above probability is approximately equal to

$$Pr\left[Z \le \frac{c_2 - 110}{15/\sqrt{n}}\right]$$ because $Pr(\bar{x} < c_1 \mid \mu = 110) \approx 0$ (See Diagram)

But $Pr(Z \le z_{0.85}) = 0.15$

or $Pr(Z \le -z_{0.15}) = 0.15$

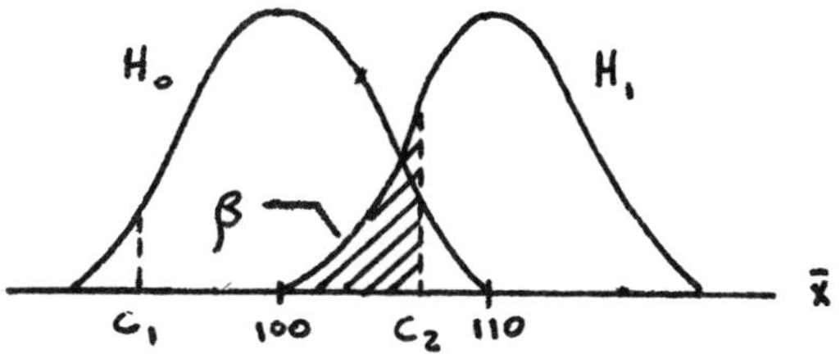

Therefore, $$\frac{(c_2 - 110)\sqrt{n}}{15} = -z_{0.15} = -1.04$$

Solving for c_2 and n, $$\frac{c_2 - 100}{c_2 - 110} = \frac{2.58}{-1.04}, \quad c_2 = 107.1$$

$$n = \left[\frac{2.58(15)}{(107.1 - 100)}\right]^2 = 29.7, \quad \text{Ans. } (n = 30)$$

Decision Rule: Reject H_0 if $\bar{x} < 92.9$ or $\bar{x} > 107.1$

7-30

The Parks and Recreation Department of a city considers a lake suitable for swimming only if the mean concentration of a particular pollutant is below 30 ppm (parts per million). A sample of 15 specimens is obtained and analyzed for its pollutant concentration, yielding a sample mean of 28 ppm with a standard deviation of 4 ppm.

Since the agency has a mandate to protect public health, it will not allow the use of the lake for swimming unless it can be convinced otherwise by the sample evidence, at a significance level of 1%. It can be assumed that pollutant concentration is normally distributed.

a) After setting up the appropriate hypotheses for this problem, determine what the Parks and Recreation Department will decide given the above sample information. Make sure that all your symbols are properly defined and that your conclusion is explicitly stated.

b) i) Explain briefly when a Type I error would be committed by the agency (in the context of this problem). Given the null hypothesis of part (a), can a Type I error be committed if the agency concludes that the lake is not suitable for swimming? Justify your answer.

ii) Explain briefly when a Type II error would be committed by the agency.

c) Assume now that the true (population) variance of pollutant concentration is known and equal to 16 $(\text{ppm})^2$. Calculate the probability of a Type II error for the test of the hypotheses of part (a) if the true mean pollutant concentration is actually equal to 27 ppm.

d) Under the assumptions of part (c), calculate the minimum sample size needed to ensure a power of 85% when the true mean pollutant concentration is actually equal to 27 ppm.

****

a) Let μ denote the mean pollutant concentration.

$H_0: \mu \geq 30$ vs. $H_1: \mu < 30$ ONE-SIDED TEST

Normal population, pop. variance unknown, small sample → t-test.

test statistic $\dfrac{\bar{X}-\mu}{s/\sqrt{n}} \sim t(n-1)$; $\bar{X}$ and s are the mean and std.-deviation of the sample, respectively.

critical region: reject H_0 if $\bar{X} \leq 30 - t_{0.01,14}\dfrac{4}{\sqrt{15}} = 30 - 2.624 \times \dfrac{4}{\sqrt{15}}$

≤ 27.29

For given data, $\bar{X} = 28 > 27.29 \Rightarrow$ DO NOT REJECT H_0; pool is unsuitable for swimming.

b) i) A Type I error would be committed if H_0 is rejected when it is actually true, i.e. if the lake is declared suitable for swimming when it really isn't.

If the lake is declared to be not suitable for swimming, then a type I error could not have occurred (since this corresponds to not rejecting H_0).

ii) A Type II error would be committed if the lake is declared to be not suitable when it actually is suitable for swimming (i.e. when μ is actually < 30 and we do not reject H_0).

c) 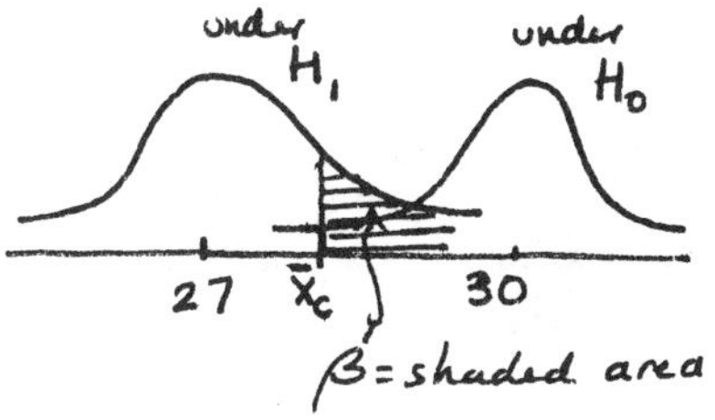

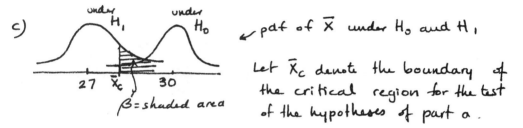

← pdf of $\bar{X}$ under H_0 and H_1

Let $\bar{X}_c$ denote the boundary of the critical region for the test of the hypotheses of part a.

Under the known pop. variance assumption, a normal test statistic will be used.

$$\bar{X}_c = 30 - z_{0.01}\frac{\sigma}{\sqrt{n}} = 30 - 2.327\,\frac{4}{\sqrt{15}} = 27.6$$

$$\beta = \text{proba. of Type II error} = \text{proba}(\text{not reject } H_0 \mid \mu = 27)$$

$$= \text{proba}(\bar{X} \geq \bar{X}_c \mid \mu = 27) = \text{proba}\left(Z \geq \frac{\bar{X}_c - 27}{4/\sqrt{15}}\right)$$

$$= 1 - \Phi\left(\frac{27.6 - 27}{4/\sqrt{15}}\right) = 1 - \Phi(0.581) = 1 - 0.719 = \underline{0.281}.$$

d) $$\text{power} = 1 - \beta = \Phi\left(\frac{\bar{X}_c - 27}{4/\sqrt{n}}\right) = \Phi\left(\frac{30 - z_{0.01}\left(\frac{4}{\sqrt{n}}\right) - 27}{4/\sqrt{n}}\right) = 0.85$$

thus $$\frac{30 - 2.327\left(\frac{4}{\sqrt{n}}\right) - 27}{4/\sqrt{n}} = \Phi^{-1}(0.85) = 1.037$$

$$30 - 2.327\left(\frac{4}{\sqrt{n}}\right) - 27 = 1.037\left(\frac{4}{\sqrt{n}}\right)$$

$$3 = (2.327 + 1.037)\left(\frac{4}{\sqrt{n}}\right) = \frac{13.456}{\sqrt{n}} \Rightarrow n = 20.12$$

use min. sample size of $\underline{\underline{21}}$ obs.

7-31

A department store sells clothing articles at an average rate of 120 articles per day with a standard deviation of 20 articles. A local newspaper advertisement was launched for 9 days and the sales averaged 125 clothing articles per day. If the type I error is $\alpha = 5\%$, determine the type II error β.

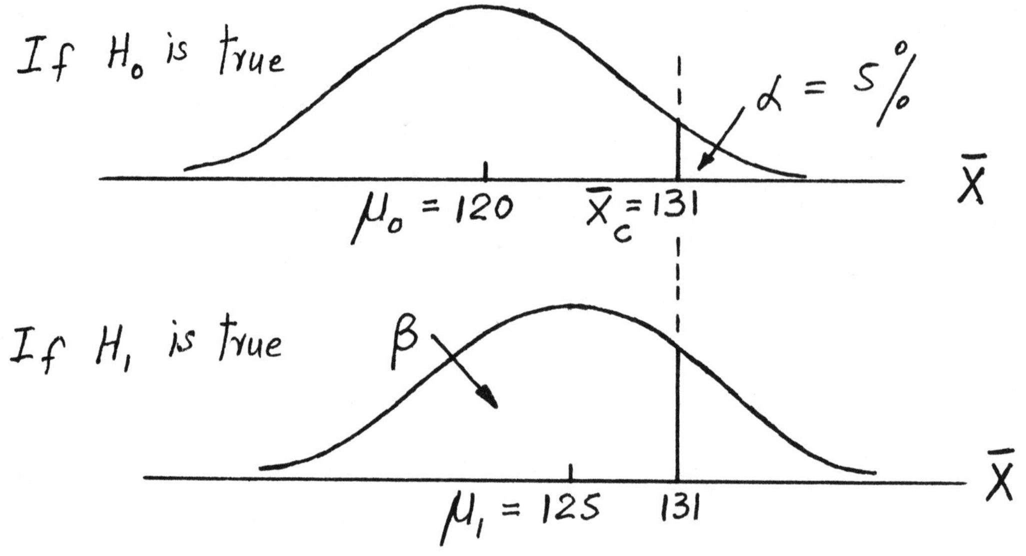

****

The Null Hypothesis $H_0 : \mu_0 = 120$
(Assume H_0 is true)

The Alternative Hypothesis $H_1 : \mu_1 = 125$

For type I error of $\alpha = 5\%$:

$$\text{critical } z_{1-\alpha} = \frac{\bar{X}_c - \mu_0}{\frac{\sigma}{\sqrt{n}}} = 1.64$$

$$\bar{X}_c = 120 + \frac{20}{\sqrt{9}} \times 1.64 \simeq 131$$

For the type II error, β:

$$\beta = P(\bar{X} < 131) = P\left(\frac{\bar{X} - \mu_1}{\frac{\sigma}{\sqrt{n}}} < \frac{131 - 125}{\frac{20}{\sqrt{9}}}\right)$$

or,

$$\beta = P(Z < 0.9) = 0.8159 \simeq 0.82$$

or, $\beta \simeq 82\%$

7-32

The changes in blood pressure of 18 patients given a drug were as follows: 7, -1, 2, 0, -5, 11, 4, 6, -2, -3, 3, 5, -1, 7, 0, 4, 8, 9. Is the assertion that this drug increases blood pressure substantiated? Show all steps using a level of significance of .01.

****

the $H_0: \mu_D = 0$ with $H_1: \mu_D > 0$; $n = 18$, $\alpha = .01$

CRITICAL REGION IS $t > t_{.01, 17} = 2.567$

the test statistic is $t = \dfrac{\bar{x} \cdot \sqrt{n}}{S_x}$, where here

x becomes d (differences).

From the data, $\bar{d} = 3$ and $S_d^2 = 20.47$

and $t = \dfrac{3\sqrt{18}}{\sqrt{20.47}} = 2.81 > 2.567$

therefore we reject Ho → the drug does increase blood-pressure.

HYPOTHESES TESTS, TWO MEANS

7-33

A company manufactures manila rope whose breaking strengths have a mean of 600 pounds and a standard deviation of 48 pounds. It is believed by a newly developed manufacturing process the mean breaking strength can be increased.

(a) Design a decision rule for rejecting the old process at the 0.01 level of significance if it is agreed to test 64 ropes.

(b) Under the decision rule adopted in (a), what is the probability of accepting the old process, when in fact the new process has increased the mean breaking strength to 620 pounds? Assume the standard deviation is still 48 pounds.

****

(a) $\alpha = 0.01$

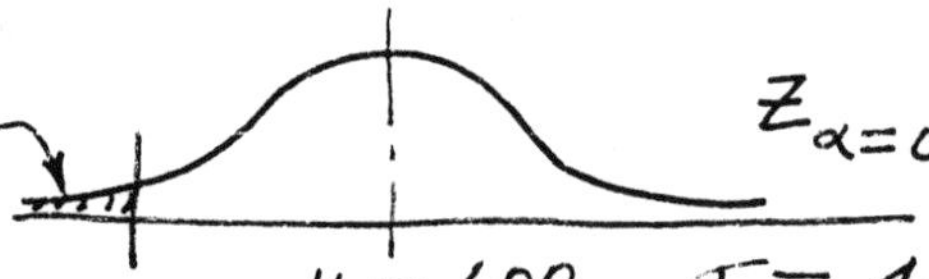

$Z_{\alpha=0.01} = -1.28$

$\mu = 600, \quad \sigma = 48$

$$\therefore -1.28 = \frac{\bar{x} - \mu}{\sigma/\sqrt{n}} = \frac{\bar{x} - 600}{48/\sqrt{64}} = \frac{\bar{x} - 600}{6}$$

$$\bar{x} = 600 - 1.28(6) = 592.32$$

$\therefore$ REJECT IF $\bar{x} < 592.32$ POUNDS

(b) $$Z_{\alpha=0.01} = \frac{\bar{x}_1 - 620}{48/\sqrt{64}} = -1.28 \qquad \bar{x}_1 = 620 - 1.28(6) = 612.32$$

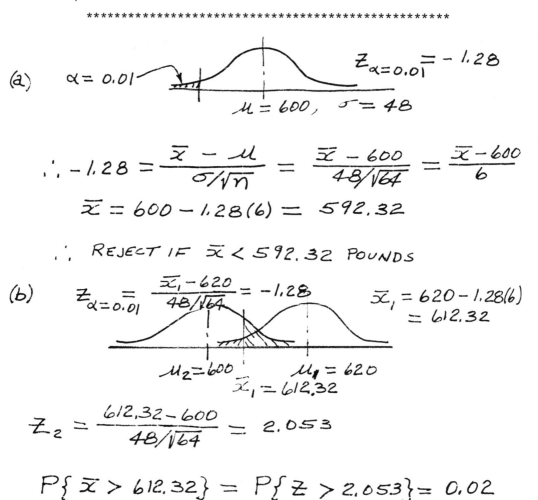

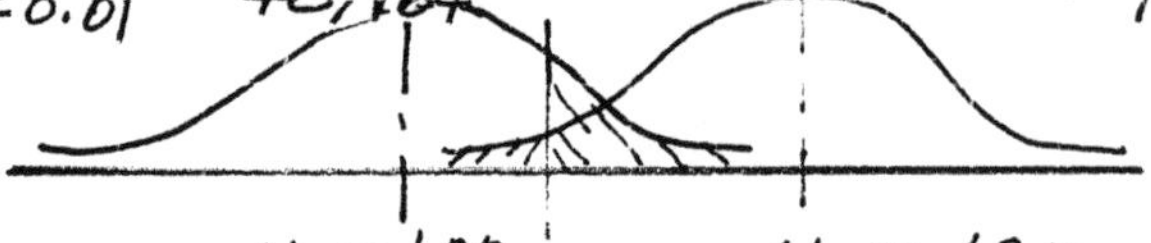

$\mu_2 = 600 \quad \mu_1 = 620$

$\bar{x}_1 = 612.32$

$$Z_2 = \frac{612.32 - 600}{48/\sqrt{64}} = 2.053$$

$$P\{\bar{x} > 612.32\} = P\{Z > 2.053\} = 0.02$$

■■■ **7-34**

Twenty laboratory mice were divided into two groups of 10 by a random selection process. Each group was fed according to a different diet. At the end of the third week the weight of each animal was recorded as below:

DIET A	5	21	7	23	11	16	13	19	12	21
DIET B	5	14	16	9	4	7	13	14	9	8

Do the data justify the conclusion that the mean weight gained on DIET A was greater than the mean weight gained on DIET B at the 5% level of significance?

****

FIRST TEST FOR EQUALITY OF VARIANCES.

$H_0 : \sigma_A^2 = \sigma_B^2$; $H_A : \sigma_A^2 \neq \sigma_B^2$; $\alpha = .05$

THIS IS A TWO-SIDED TEST WITH THE F-DISTRIBUTION.

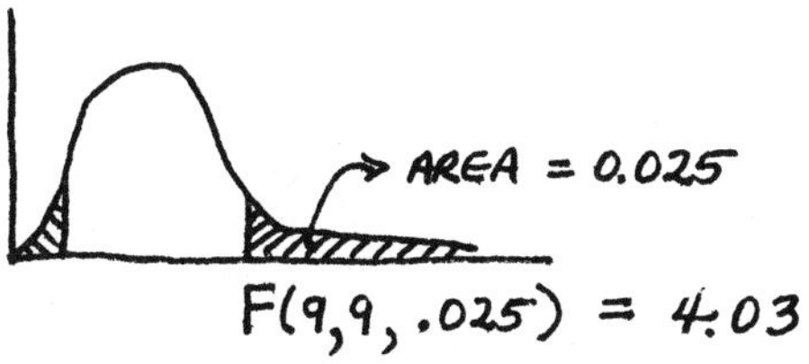

$F^* = S_A^2 / S_B^2 = 2.265 < 4.03$

THUS, ASSUME $\sigma_A = \sigma_B$.

AREA = 0.025

$F(9,9,.025) = 4.03$

$n_A = 10$, $\overline{X}_A = 14.8$, $S_A = 6.20$
$n_B = 10$, $\overline{X}_B = 9.9$, $S_B = 4.12$

NOW TEST THE DIFFERENCES IN THE MEANS.

$H_0 : \mu_A \leq \mu_B$; $H_A : \mu_A > \mu_B$; $\alpha = .05$; $df = 18 (= 10+10-2)$

THIS IS A ONE-SIDED TEST WITH THE t-DISTRIBUTION.

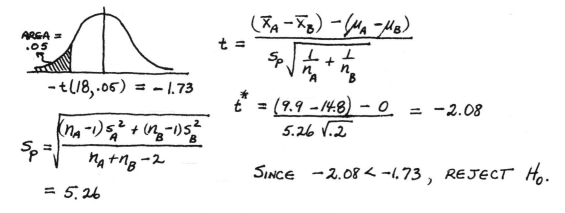

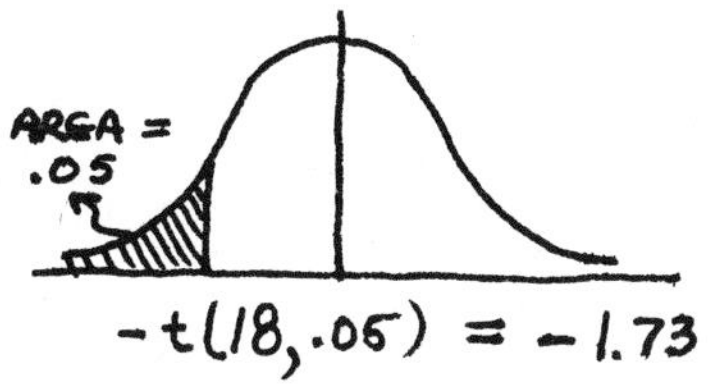

AREA = .05

$-t(18,.05) = -1.73$

$$t = \frac{(\overline{X}_A - \overline{X}_B) - (\mu_A - \mu_B)}{S_P \sqrt{\frac{1}{n_A} + \frac{1}{n_B}}}$$

$$t^* = \frac{(9.9 - 14.8) - 0}{5.26 \sqrt{.2}} = -2.08$$

$$S_P = \sqrt{\frac{(n_A - 1) S_A^2 + (n_B - 1) S_B^2}{n_A + n_B - 2}}$$

$$= 5.26$$

SINCE $-2.08 < -1.73$, REJECT H_0 .

7-35

The productivity of men and women at an assembly plant is being investigated. A sample of 36 men and 36 women produced the following results:

	Men	Women
Total Number of Parts Assembled in 1 Hour:	2196	2088

Past experience suggests equal standard deviations of 5 parts/hr for both men and women. Test the hypothesis, at the 5% level of significance, that productivity is independent of the sex of the worker.

****

Let X_1 = Hourly output of male employees with mean μ_1
X_2 = Hourly output of female employees with mean μ_2

$H_0: \mu_1 = \mu_2$ or $\mu_1 - \mu_2 = 0$
$H_1: \mu_1 > \mu_2$ or $\mu_1 - \mu_2 > 0$

Note, the alternate hypothesis could just as well have been $H_1: \mu_1 \neq \mu_2$ or $\mu_1 - \mu_2 \neq 0$ although data suggests other H_1.

The samples are large enough to suggest that

$$\bar{x}_1 \sim N(\mu_1, \sigma_1^2/n_1) \quad \text{and} \quad \bar{x}_2 \sim N(\mu_2, \sigma_2^2/n_2)$$

This is a one-tail test with critical region obtained as follows:

Under H_0

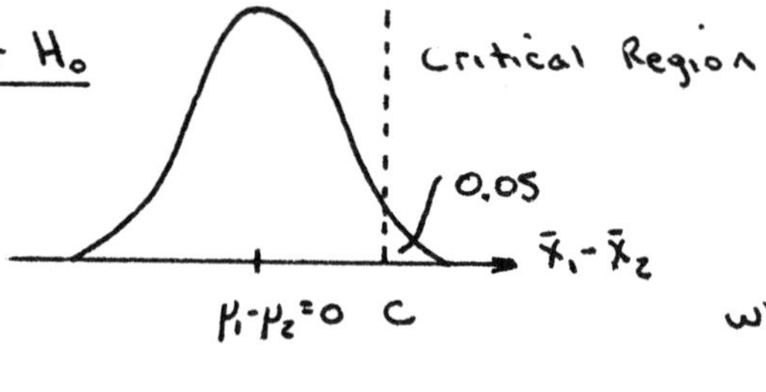

$$\Pr(\bar{x}_1 - \bar{x}_2 > c \mid \mu_1 - \mu_2 = 0) = 0.05$$

$$\Pr\left[Z > \frac{c - \mu_{\bar{x}_1 - \bar{x}_2}}{\sigma_{\bar{x}_1 - \bar{x}_2}}\right] = 0.05$$

where $\mu_{\bar{x}_1 - \bar{x}_2} = \mu_1 - \mu_2 = 0$

$$\sigma_{\bar{x}_1 - \bar{x}_2} = \sqrt{\frac{\sigma_1^2}{n_1} + \frac{\sigma_2^2}{n_2}}$$

$$= \sqrt{\frac{25}{36} + \frac{25}{36}} = 1.179$$

Since $\Pr(Z > z_{0.05}) = 0.05$

It follows that, $\dfrac{c - 0}{1.179} = z_{0.05} = 1.645$

$c = 1.94$

Decision Rule: Reject H_0 if $\bar{x}_1 - \bar{x}_2 > 1.94$

$\bar{x}_1 = \dfrac{2196}{36} = 61$, $\bar{x}_2 = 58$

Since $\bar{x}_1 - \bar{x}_2 = 61 - 58 > 1.94$ we reject H_0

7-36

An engineering society wishes to determine if the mean income of practicing electrical engineers exceeds that of civil engineers. Two independent random samples provided the following data:

	Electrical Engineers	Civil Engineers
sample size	76	58
mean	\$37,246	\$36,412
standard deviation	\$8,371	\$8,856

a) In comparing the mean income μ_1 and μ_2 of electrical and civil engineers respectively, the following hypotheses were formulated: $H_0: \mu_1 - \mu_2 \geq 0$ against $H_1: \mu_1 - \mu_2 < 0$. [Give a brief explanation (in words) of the underlying logic (for instance, which hypothesis was given the benefit of the doubt?)].

b) Use the above sample information to perform the test stated in question (a). Use a significance level of 5%.

****

a) The hypothesis that electrical engineers have higher incomes than civil engineers was given the benefit of the doubt.

b) In this problem, we use the large-sample z-test.

test statistic: $$\frac{\bar{X}_1 - \bar{X}_2}{\sqrt{\frac{S_1^2}{n_1} + \frac{S_2^2}{n_2}}} \underset{\text{approx.}}{\sim} N(0,1)$$

where $\bar{X}_i, S_i$ denote the sample mean and std. deviation, respect., for sample i, $i=1,2$.

critical region: reject H_0 if

$$\bar{x}_1 - \bar{x}_2 < -z_{0.05}\sqrt{\frac{s_1^2}{n_1} + \frac{s_2^2}{n_2}}$$

$$< -1.645\sqrt{\frac{(8{,}371)^2}{76} + \frac{(8{,}856)^2}{58}}$$

$$< -2480.76$$

For given data,

$$\bar{x}_1 - \bar{x}_2 = 37{,}246 - 36{,}412 = 834 > -2480.76$$

⇒ DO NOT REJECT H_0;

EE's have higher incomes than CE's.

7-37

A certain corporation wishes to compare the mean salaries of its workers at two of its plants located in two different towns. A random sample of $n_1 = 20$ workers at the first plant showed an average earnings of $\overline{X}_1 = \$800$ per week per worker and a standard deviation of $\sigma_1 = \$200$; whereas, a sample of $n_2 = 30$ workers at the second plant indicated an average salary of $\overline{X}_2 = \$900$ per week per worker and a standard deviation of $\sigma_2 = \$270$. Determine on the basis of a two sided test at $\alpha = 5\%$ level if there is a significant difference between the mean salaries of workers at the two plants. Assume the mean salaries are normally distributed at the both plants.

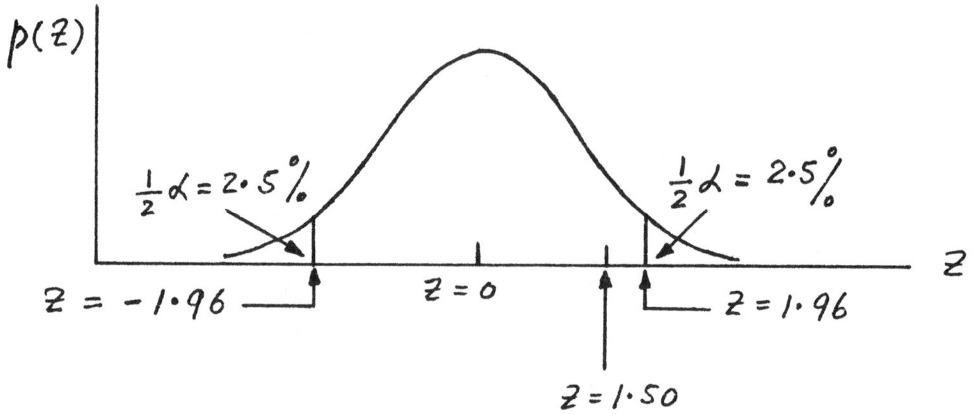

****

The Null Hypothesis $H_0: \mu_1 = \mu_2$ (Assume H_0 is true)

The Alternative Hypothesis $H_1: \mu_1 \neq \mu_2$

The two critical regions at $\frac{1}{2}\alpha = 2.5\%$ are shown on either end of the standar normal curve for the probability distribution as drawn in the above figure. The corresponding values of the standard normal variable for the rejection probability at $\frac{1}{2}\alpha$ on the lower and upper tails of the standard normal distribution are, respectively, $z = -1.96$ and

$z = 1.96$. The standard normal variable for the test statistics is

$$Z = \frac{\bar{X}_2 - \bar{X}_1}{\sqrt{\frac{\sigma_1^2}{n_1} + \frac{\sigma_2^2}{n_2}}} = \frac{900 - 800}{\sqrt{\frac{(200)^2}{20} + \frac{(270)^2}{30}}} = 1.50$$

Hence, it can be concluded that there is no significant difference in the salaries of workers at both the plants.

7-38

Dimensions on two samples of rheostat knobs gave the following data:

$n_1 = 5, \bar{x}_1 = 137.8, S_1 = 3.31$

$n_2 = 5, \bar{x}_2 = 143.0, S_2 = 2.76$

Is there evidence of a difference between the two samples (use $\alpha = .05$)?

****

Assuming $\sigma_1^2 = \sigma_2^2$, we can set up the following test:

$H_0: \mu_1 = \mu_2$ and $H_1: \mu_1 \neq \mu_2$

with $n_1 = n_2 = 5$ and $\alpha = .05$

COMPUTE $S_p^2 = \frac{S_1^2 + S_2^2}{2} = \frac{(3.31)^2 + (2.76)^2}{2} = 9.29$

the test statistic is $t = \frac{|\bar{X}_1 - \bar{X}_2|}{\sqrt{S_p^2\left(\frac{1}{n_1} + \frac{1}{n_2}\right)}}$

with the critical region being $|t| > t_{.025,8} = 2.306$

then computed $t = \frac{5.2}{\sqrt{9.287\left(\frac{1}{5} + \frac{1}{5}\right)}} = 2.698$

$\therefore$ Reject H_0 $\rightarrow$ the samples came from different populations.

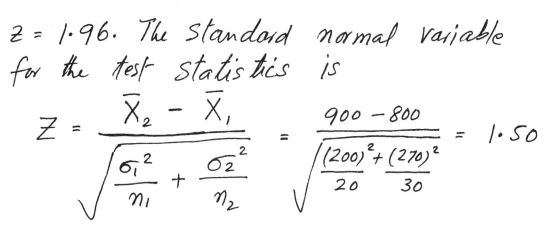

7-39

At the Yeastside Westside Bakery, two packing lines box doughnuts. Each box of packaged doughnuts is weighed before it is released to make certain it passes a minimum weight standard. Slim Chance boxes doughnuts on Line A. Ima Glutton, on Line B, snitches a doughnut from time to time and puts in its place a piece of metal of comparable weight. Abel Baker, the owner, suspects something following customer complaints and concludes that some sort of test is kneaded before doughnut sales go in the hole. Consequently, the following sample weight measurements were taken for boxes of doughnuts coming off each line. Are the variances and the means of the samples significantly different at the 5% significance level?

Line A: 14, 13, 17, 14, 15, 17 (weights in ounces)
Line B: 17, 18, 20, 19, 17

****

The variances s_A^2 and s_B^2 estimated for Lines A and B should be compared first using the F test:

$$s_A^2 = \frac{(6)(\Sigma\,\text{weights}^2) - (\Sigma\,\text{weights})^2}{(6)(5)} = \frac{(6)(1{,}364) - (90)^2}{(6)(5)} = 2.8$$

$$s_B^2 = \frac{(5)(1{,}663) - (91)^2}{(5)(4)} = 1.7$$

$$F = \frac{s_A^2}{s_B^2} = \frac{2.8}{1.7} = 1.65 < 6.26 \quad \left(\text{Table F at 5\% significance, 5 degrees of freedom for numerator, 4 for denominator}\right)$$

No significant difference in variances on Lines A and B. t test may be run to compare means $\bar{x}_A$ and $\bar{x}_B$:

$$\bar{x}_A = \frac{90}{6} = 15.0 \qquad \bar{x}_B = \frac{91}{5} = 18.2$$

$$t = \frac{|15.0 - 18.2|}{\sqrt{\dfrac{(2.8)(5) + (1.7)(4)}{9} \cdot \dfrac{11}{30}}} = 3.48 > 2.262 \quad \left(\text{Table t at 5\% level and 9 degrees of freedom}\right)$$

This shows a significant difference in average weights on Lines A and B.

8
PROPORTIONS

POINT ESTIMATES

8-1

GALLUP POLLED 785 ADULTS AND
FOUND THAT 74% DID NOT THINK THAT THE USE OF MARIJUANA SHOULD BE LEGALIZED.

HOW BIG OF A SAMPLE SHOULD THEY HAVE TAKEN TO GET A STANDARD ERROR OF THE ESTIMATE OF 1% ? IS THERE ANY BIAS IN THEIR SURVEY ?

****

TO ANSWER THE SECOND QUESTION FIRST, MANY WOULD FEEL THAT THERE PROBABLY WAS BIAS IN THE SURVEY FOR THE SIMPLE REASON THAT ONLY ADULTS WERE CONTACTED. CERTAINLY A BETTER SURVEY WOULD HAVE INCLUDED ALL AGE GROUPS AFFECTED BY THE QUESTION.

$\alpha = 0.05 \quad Z_{\alpha/2} = 1.96 \quad \hat{p} = 0.74 \quad \hat{q} = .26$

FOR $E = 0.01$, $\quad N = pq\left[\frac{Z_{\alpha/2}}{E}\right]^2$

$$= (.74)(.26)\left[\frac{1.96}{.01}\right]^2 = 7392$$

8-2

The process of taking an opinion poll is that of taking a random sample. Most poll results are expressed as the percentage of persons giving a certain answer, and are used to estimate the percentage of persons in the population who hold a particular opinion. If a poll of 700 persons showed that 343 answered "no", what is the point estimate of the percentage of the population which feels that "no" is the wrong answer?

****

$$\hat{p} = 1 - \frac{343}{700} = 0.510$$

8-3

Five per cent of the TV's produced by a certain manufacturer have a defective picture tube. Find the probability that more than 4% of a random sample of 450 TV's will have a defective picture tube.

****

Let π = % defective in population = 0.05
p = % defective in sample

$p \sim N(\mu_p, \sigma_p)$ where $\mu_p = \pi = 0.05$

$$\sigma_p = \sqrt{\frac{\pi(1-\pi)}{n}} = \sqrt{\frac{(0.05)(0.95)}{450}}$$

$$\begin{aligned}
Pr(p > 0.04) &= Pr\left[\frac{p-\mu_p}{\sigma_p} > \frac{0.04-0.05}{\sqrt{\frac{0.0475}{450}}}\right] \\
&= Pr(Z > -0.973) \\
&= 1 - Pr(Z \leq -0.973) \\
&= 1 - F_Z(-0.973) \\
&= 1 - \{1 - F_Z(0.973)\} \\
&= F_Z(0.973) \\
&= 0.835
\end{aligned}$$

Note, $p \sim N(\mu_p, \sigma_p)$ is based on the Normal approximation to the binomial distribution.

8-4

Suppose that 10% of the tools produced in a certain manufacturing process turns out to be defective. Find the probability that in a sample of 10 tools selected at random, exactly 2 will be defective by using:

(a) The binomial distribution

(b) The Poisson approximation to the binomial distribution

(c) Normal approximation to the Poisson distribution, e.g., determine P(1.5 < x < 2.5).

****

(a) $p = 0.10, \quad n = 10, \quad k = 2 \qquad P\{k\} = \binom{n}{k} p^k (1-p)^{n-k}$

$$P\{k=2\} = \binom{10}{2}(0.10)^2(0.90)^8 = 0.194$$

(b) $np = 10(0.10) = 1.0, \quad P\{k\} = \dfrac{(np)^k e^{-np}}{k!}$

$$P\{k=2\} = \frac{(1.0)^2 e^{-1.0}}{2!} = 0.184$$

(c) $P\{1.5 < x < 2.5\} = P\{z_1 < z < z_2\}$

$$z_1 = \frac{x_1 - \mu}{\sigma} = \frac{1.5 - 10(0.10)}{\sqrt{10(0.10)}} = 0.5$$

$$z_2 = \frac{x_2 - \mu}{\sigma} = \frac{2.5 - 10(0.10)}{\sqrt{10(0.10)}} = 1.5$$

$$P\{0.5 < z < 1.5\} = P\{z < 1.5\} - P\{z < 0.5\}$$

$$= 0.9332 - 0.6915 = 0.2417$$

NOTE: n is too small for a good approximation.

INTERVAL ESTIMATES

8-5

A RANDOM SURVEY OF ENGINEERING STUDENTS AT A UNIVERSITY SHOWED THAT 317 OUT OF 591 READ THE CAMPUS NEWSPAPER MORE THAT ONCE A WEEK.

WHAT IS THE ERROR IN THIS ESTIMATE OF THE TRUE PERCENTAGE AT A 99% CONFIDENCE LEVEL ? AT A 95% CONFIDENCE LEVEL ?

****

$$\hat{p} = \frac{317}{591} = 0.536$$

99% C.I. $\quad E = Z\sqrt{\frac{\hat{p}\,\hat{q}}{N}} = Z\sqrt{\frac{(.536)(1-.536)}{591}}$

$= 0.02\,Z$

$$Z_{.99} = 2.576$$

$$E = 0.02\,(2.576) = 0.0528$$

$\therefore \quad p = 0.536 \pm 0.0528$

95% C.I. $\quad Z_{.95} = 1.96$

$$E = 0.02\,(1.96) = 0.0392$$

$\therefore \quad p = 0.536 \pm 0.0392$

8-6

Sixty per cent of a sample of 100 voters favor a certain bill. Find a 99% confidence interval for the true proportion of voters favoring the bill.

****

Let π = True (unknown) proportion favoring bill

The sample statistic p, i.e. sample proportion is approximately Normally distributed with std. deviation

$$\sigma_p = \sqrt{\frac{\pi(1-\pi)}{n}} \approx \sqrt{\frac{p(1-p)}{n}} \quad \text{for large } n$$

$$99\% \text{ C.I. for } \pi = p \pm z_{0.005}\,\sigma_p$$

$$= 0.6 \pm 2.58\sqrt{\frac{0.6(0.4)}{100}}$$

$$= 0.6 \pm 0.126$$

8-7

A bank randomly selected 150 checking account customers and found that 68 of them also had savings accounts at the same bank. Construct a 90% confidence interval estimate for the true proportion of checking account customers that also have savings accounts.

****

$n = 150$ THE ESTIMATE FOR THE PROPORTION P IS x/n —

$x = 68$

$\alpha = 0.1$

$$\hat{p} = x/n = 68/150 = 0.453 \;;\; \hat{q} = 1 - \hat{p} = 0.547$$

THIS IS A TWO-SIDED TEST WITH A NORMAL C.D.F.

$$z = \frac{\hat{p} - P}{\sqrt{\frac{\hat{p}\hat{q}}{n}}} \;;\quad \hat{p} - z_{.05}\sqrt{\frac{\hat{p}\hat{q}}{n}} \le p \le \hat{p} + z_{.05}\sqrt{\frac{\hat{p}\hat{q}}{n}}$$

$$.453 - 1.645(.041) \le p \le .453 + 1.645(.041)$$

$$\pm z_{.05} = \pm 1.645 \qquad .386 \le p \le .520$$

8-8

Twenty five out of 250 items produced by process I are defective, and 15 out of 180 produced by process II are defective. Construct a) 95% and b) 99% confidence intervals for the difference in proportion of defective items produced by the two processes.

****

Let P_1 = Sample proportion from process I
P_2 = Sample proportion from process II

i.e. P_1 and P_2 are point estimates of the unknown process I and II per cent defectives, π_1 and π_2, respectively.

Based on the Normal approximation to the binomial distribution,

$$P_1 \sim N(\mu_{P_1}, \sigma_{P_1}) \quad \text{and} \quad P_2 \sim N(\mu_{P_2}, \sigma_{P_2})$$

where

$$\mu_{P_1} = \pi_1 \qquad \mu_{P_2} = \pi_2$$

$$\sigma^2_{P_1} = \frac{\pi_1(1-\pi_1)}{n_1} \qquad \sigma^2_{P_2} = \frac{\pi_2(1-\pi_2)}{n_2}$$

The estimator of $\pi_1 - \pi_2$ is $P_1 - P_2$ which is also approximately Normally distributed

$$P_1 - P_2 \sim N(\mu_{P_1-P_2}, \sigma_{P_1-P_2}) \quad \text{where} \quad \mu_{P_1-P_2} = \mu_{P_1} - \mu_{P_2} = \pi_1 - \pi_2$$

$$\text{and} \quad \sigma^2_{P_1-P_2} = \sigma^2_{P_1} + \sigma^2_{P_2} = \frac{\pi_1(1-\pi_1)}{n_1} + \frac{\pi_2(1-\pi_2)}{n_2}$$

The $100(1-\alpha)\%$ confidence interval for $\pi_1 - \pi_2$ is given by

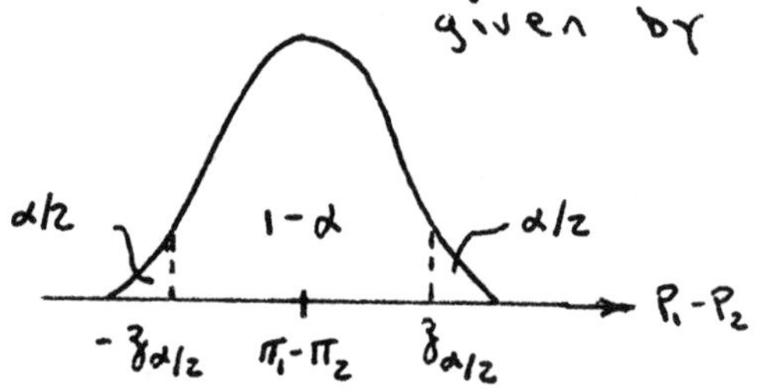

$$P_1 - P_2 \pm z_{\alpha/2}\, \sigma_{P_1-P_2}$$

$$= P_1 - P_2 \pm z_{\alpha/2} \sqrt{\frac{\pi_1(1-\pi_1)}{n_1} + \frac{\pi_2(1-\pi_2)}{n_2}}$$

Since π_1 and π_2 are unknown, p_1 and p_2 are used instead. Hence,

$$100(1-\alpha)\%\ \text{C.I. for } \pi_1-\pi_2 = p_1-p_2 \pm z_{\alpha/2}\sqrt{\frac{p_1(1-p_1)}{n_1}+\frac{p_2(1-p_2)}{n_2}}$$

In this problem, $p_1 = \frac{25}{250} = 0.1$ and $p_2 = \frac{15}{180} = 0.083$

a) 95% C.I., $\alpha = 0.05$, $z_{\alpha/2} = z_{0.025} = 1.96$

$$0.1-0.083 \pm 1.96\sqrt{\frac{(0.1)(0.9)}{250}+\frac{(0.083)(0.917)}{180}} = 0.017 \pm 0.0548$$

b) 99% C.I., $\alpha = 0.01$, $z_{\alpha/2} = z_{0.005} = 2.58$

$$0.1-0.083 \pm 2.58\sqrt{\frac{(0.1)(0.9)}{250}+\frac{(0.083)(0.917)}{180}} = 0.017 \pm 0.0721$$

8-9

The first 200 babies born in Cincinnati in January were 120 boys and 80 girls. Estimate by an 80% confidence interval the true proportion of male births in Cincinnati.

****

$$\hat{p} = 120/200 = 0.60$$

$$\hat{\sigma}_p = \sqrt{\frac{\hat{p}\hat{q}}{n}} = \sqrt{\frac{(.6)(.4)}{200}} = .0346$$

and $\hat{p} \pm Z_{.10}\,\hat{\sigma}_p = 0.60 \pm (1.282)(.0346)$

$$= 0.60 \pm .044 = [.556, .644]$$

$$\therefore P(.556 < p < .644) = 0.80$$

8-10

What is the largest sample necessary in order to establish a 95% confidence interval on the true value of a proportion if the maximum allowable error is plus or minus 2%?

****

USING THE LARGE SAMPLE APPROXIMATION,

$$0.02 = z_{\alpha/2}\sqrt{\frac{p(1-p)}{n}} = 1.96\sqrt{\frac{p(1-p)}{n}}$$

SINCE $p(1-p)$ IS A MAXIMUM AT $p = 0.5$, WE USE $p = 0.5$ TO FIND THE LARGEST NECESSARY SAMPLE.

$$1.96\sqrt{\frac{0.25}{n}} = 0.02$$

$$n = 2401$$

8-11

In a random sample of 100 diodes, 28 are found to fall in a certain voltage category. Construct a 99% confidence interval for the proportion of diodes in that category.

****

Estimate of p is $\hat{p} = 28/100 = 0.28$

Estimate of σ_p is $\hat{\sigma}_p = \sqrt{\frac{\hat{p}(1-\hat{p})}{n}} = \sqrt{\frac{(.28)(.72)}{100}} = .0449$

and $\hat{p} \pm Z_{.005}\,\hat{\sigma}_p = .28 \pm (2.576)(.0449)$

$= .28 \pm .115$

$\therefore P(.165 < p < .395) = 0.99$

8-12

A campaign manager for a political candidate wishes to estimate the per cent of eligible voters who favor his candidate. He is willing to accept an error of no more than plus or minus five per cent at a confidence level of 95%. Find the required sample size if

a) he has no knowledge whatsoever about the true per cent of voters favoring his candidate.

b) he has reason to believe the true per cent of voters favoring his candidate is in the neighborhood of 30%.

****

$\alpha = 0.05$

$z_{\alpha/2} = z_{0.025} = 1.96$

$$n = \pi(1-\pi)\left[\frac{z_{\alpha/2}}{\epsilon_{MAX}}\right]^2$$

where π = True % of voters favoring candidate

ϵ_{MAX} = Maximum allowable error (0.05)

a) With no knowledge of π, we can set $\pi = 0.5$ to obtain the maximum possible sample size

$$n = 0.5(1-0.5)\left(\frac{1.96}{0.05}\right)^2 = 384 \text{ nearest integer}$$

b) Using $\pi = 0.3$ to determine the sample size

$$n = 0.3(1-0.3)\left(\frac{1.96}{0.05}\right)^2 = 323 \text{ nearest integer}$$

HYPOTHESES TESTS: ONE PROPORTION

8-13

Two friends, Wayne and Henry, worked together one summer and played 120 sets of tennis during that time. If Wayne won 70 of the 120 sets, test the hypothesis that Wayne and Henry are equally skilled at tennis. Use a level of significance of 5%.

****

$H_0: p = 0.5$

$H_1: p \neq 0.5$

TEST STATISTIC = z

CRITICAL VALUE = 1.96

$\alpha = 0.05$

WITHOUT THE CONTINUITY CORRECTION

$$z = \frac{x - np_0}{\sqrt{np_0(1-p_0)}} = \frac{70-60}{\sqrt{60(0.5)}} = 1.83$$

$1.83 < 1.96$, AND CANNOT REJECT H_0

WITH THE CONTINUITY CORRECTION

$$z = \frac{x - 1/2 - np_0}{\sqrt{np_0(1-p_0)}} = 1.735$$

AND AGAIN, CAN NOT REJECT H_0

8-14

Suppose when sampling from a binomial population, we wish to test

$H_0: \pi = 0.4$ against $H_1: \pi = 0.6$

where π is the mean of the binomial population. Using the sample proportion $p = \frac{X}{n}$ as the test statistic, where X is binomially distributed with parameters n and π, find the sample size and critical value of p in order that

α = Pr(Type I error) = 0.01 and β = Pr(Type II error) = 0.10

****

$Pr(p > C \mid \pi = 0.4) = 0.01$ where C is the critical value

$Pr(X > nC \mid \pi = 0.4) = 0.01$ since $p = X/n$

Using the Normal approximation to the binomial distribution

$X \sim b(n, \pi)$ with $\mu_X = n\pi$ & $\sigma_X^2 = n\pi(1-\pi)$

Approximation: $X \sim N(\mu_X, \sigma_X)$ with $\mu_X = n\pi$ & $\sigma_X^2 = n\pi(1-\pi)$

$$Pr(X > nC \mid \pi = 0.4) = Pr\left[\frac{X - \mu_X}{\sigma_X} > \frac{nC - 0.4n}{\sqrt{n(0.4)(0.6)}}\right] = 0.01$$

$$Pr\left[Z > \frac{nC - 0.4n}{\sqrt{n(0.24)}}\right] = 0.01 \Longrightarrow \frac{nC - 0.4n}{\sqrt{n(0.24)}} = z_{0.01}$$

$Pr(p \le C \mid \pi = 0.6) = 0.10$

$Pr(X \le nC \mid \pi = 0.6) = 0.10$

$$Pr\left[\frac{X - \mu_X}{\sigma_X} \le \frac{nC - 0.6n}{\sqrt{n(0.6)(0.4)}}\right] = Pr\left[Z \le \frac{nC - 0.6n}{\sqrt{n(0.24)}}\right] = 0.10$$

$$\Longrightarrow \frac{nC - 0.6n}{\sqrt{n(0.24)}} = z_{0.90} = -z_{0.10}$$

Solving for n and C,

$$\frac{C - 0.4}{C - 0.6} = \frac{z_{0.01}}{-z_{0.10}} = \frac{2.33}{-1.28} \quad \text{which gives } C = 0.529$$

Using either equation above gives $n = 78.3$ Ans. $n = 79$

8-15

A quality controller suggests that 10% of the electric fuses in a large batch are defective. To verify his hypothesis, he picks a sample of 400 fuses and tests each fuse in the sample. He observed that 50 fuses were defective. Does the test result violate the hypothesis at the 5% level of significance?

****

Suppose p is the true proportion of bad fuses.

The null hypothesis is

$$H_0 : p = 0.1$$

ad the alternative hypothesis is

$$H_1 : p > 0.1$$

If there are N defective fuses in a sample of m, the defective proportion in the sample $\bar{p} = \frac{N}{m}$

Now if σ is the std. deviation of $\bar{p}$, the random variable

$$z = \frac{\bar{p} - p}{\sigma}$$

has zero mean and unity std. deviation. Assume normal distribution for z. From tables,

$$P(z > z_0) = 0.05$$

gives

$$z_0 = 1.645$$

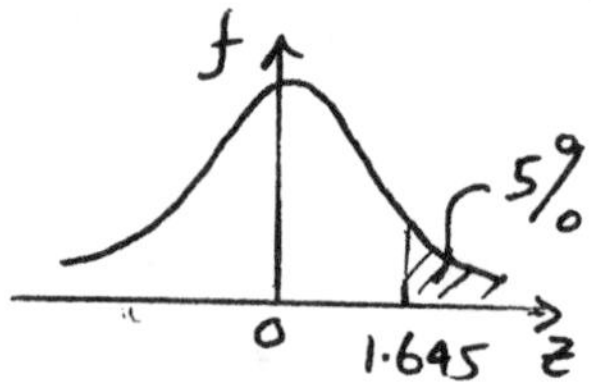

The implication here is that an experimental deviation of greater than 1.645 is highly improbable (at 5%) if H_0 is true.

Now let us check what the experimental value of z is.

If H_0 is true, $p = 0.1$ and

$$\sigma^2 = \frac{p(1-p)}{m} = \frac{0.1(1-0.1)}{400}$$

{ NOTE: N is binomial with Variance $mp(1-p)$. Hence $\frac{N}{m}$ has a variance $\frac{p(1-p)}{m}$ }

or, $\sigma = 0.015$

The experimental value of $\bar{p}$ is $= \frac{50}{400} = 0.125$

The corresponding value of z is

$$z = \frac{0.125 - 0.1}{0.015} = 1.667$$

This value is greater than the allowed upper limit 1.645. It can be concluded that the test violates the hypothesis H_0 at the 5% level of significance.

HYPOTHESES TESTS: TWO PROPORTIONS

8-16

During a work sampling study, an IE Department classified workers as productive or nonproductive. During week 10 of the study, the observed fraction productive was 0.72 of 50 workers. During week 11, the observed fraction productive was 0.65 of 60 workers. Assuming that the observations comprise a random sample, is there sufficient evidence to believe that there has been a reduction in the fraction of productive workers in the work force? Use a significance level of 5%.

****

$H_0: p_1 = p_2$ TEST STATISTIC $= Z$

CRITICAL VALUE $= 1.645$

$H_1: p_1 > p_2$ $\alpha = 0.05$

$$\hat{p} = \frac{36+39}{110} = 0.682$$

$$Z = \frac{0.72 - 0.65}{\sqrt{(0.682)(0.318)\left(\frac{1}{50}+\frac{1}{60}\right)}} = 0.785$$

$0.785 < 1.645$, AND CANNOT REJECT H_0

8-17

Two groups of 50 patients each took part in an experiment in which one group of 50 received pills containing an anti-allergy drug and the other group of 50 received a placebo (a pill containing no drug). In the group given the drug 15 exhibited allergic symptoms while in the group containing the placebo 24 exhibited such symptoms. Is the evidence sufficient to conclude at the .05 levels that the drug is effective in reducing these symptoms? Show all steps in the analysis.

****

$H_0: p_1 = p_2$ with $H_1: p_1 > p_2$ with $n_1 = n_2 = 50$, $\alpha = .05$

the test statistic is $Z = \dfrac{\hat{p}_1 - \hat{p}_2}{\sqrt{\hat{p}(1-\hat{p})\left(\frac{1}{n_1}+\frac{1}{n_2}\right)}}$

where $\hat{p}_1 = 24/50 = .48$, $\hat{p}_2 = 15/50 = .30$, $\hat{p} = \dfrac{\hat{p}_1 + \hat{p}_2}{2} = .39$

the critical region is $Z > Z_{.05} = 1.645$

then $Z = \dfrac{.18}{\sqrt{(.39)(.61)(.04)}} = 1.845 > 1.645$

$\therefore$ WE REJECT H_0 $\rightarrow$ the drug is effective.

8-18

A survey revealed that a household product is used by 128 out of 400 people interviewed in a city where the product was widely advertised, but only 115 out of 500 people used it in a city with no advertising. Determine whether or not advertising is effective at the 95% confidence level.

****

A = CITY WITH ADVERTISING ; P_A = % USING PRODUCT IN A
B = CITY WITHOUT ADVERTISING ; P_B = % USING PRODUCT IN B

$N_A = 400$, $N_B = 500$, $P_A = 128/400$, $P_B = 115/500$

H_0 : $P_A = P_B = P$ (ADVERTISING NOT EFFECTIVE)
H_A : $P_A > P_B$

THIS IS A ONE-SIDED TEST WITH A NORMAL C.D.F.
SINCE H_0 IS $P_A = P_B$, THE BEST ESTIMATE FOR THE TRUE P IS

$$\hat{p} = \frac{128 + 115}{400 + 500} = 0.27 \quad , \quad \hat{q} = 1 - \hat{p} = 0.73$$

$$Z = \frac{P_A - P_B}{\sqrt{\hat{p}\hat{q}\left(\frac{1}{N_A} + \frac{1}{N_B}\right)}} = 3.02$$

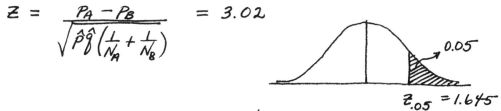

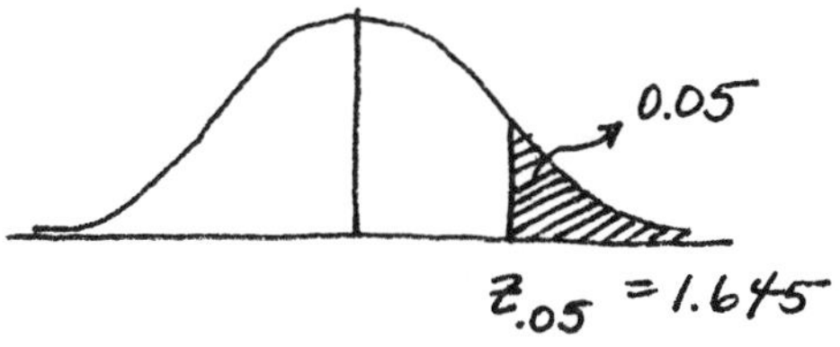

SINCE $3.02 > 1.645$, REJECT H_0 .

8-19

TWO GROUPS OF PATIENTS (80 EACH) WERE USED TO TEST THE EFFECTIVENESS OF A NEW DRUG. THE FIRST GROUP WAS GIVEN THE DRUG AND 57 SHOWED SIGNS OF IMPROVEMENT. THE SECOND GROUP WAS GIVEN A PLACEBO (A PILL NOT CONTAINING ANY DRUG) AND 39 SHOWED IMPROVEMENT. CAN WE CONCLUDE THAT THE DRUG WAS EFFECTIVE AT A SIGNIFICANCE LEVEL OF

0.01 ? 0.05 ? 0.10 ?

****

$$\hat{P}_1 = \frac{57}{80} = 0.7125 \qquad \hat{P}_2 = \frac{39}{80} = 0.4875$$

$$Z = \frac{\hat{P}_1 - \hat{P}_2}{\sqrt{\frac{\hat{P}_1 \hat{q}_1}{N_1} + \frac{\hat{P}_2 \hat{q}_2}{N_2}}} = 2.9846$$

$H_0 : p_1 > p_2 \qquad H_1 : p_1 = p_2$

ACCEPT IF $Z > Z_\alpha$

$\alpha = 0.01$	2.9846 :	2.326	ACCEPT
$\alpha = 0.05$	" :	1.645	"
$\alpha = 0.10$	" :	1.282	"

THE DRUG IS MORE EFFECTIVE AT ALL THESE α LEVELS

8-20

The following is actual data for a certain college's students who took the Engineering Intern Examination on April 1982

Department	Major	Passed	Failed
Civil	Civil	9	10
	Environmental	4	2
Industrial	Industrial	2	2
	EMCS	6	2
Electrical	Electrical	37	10
Mechanical	Mechanical	18	5
		76	31

a) Does the major make a difference when it comes to preparation for the EI exam? Use a 5% level of significance.

b) If you are told that the proportion of failing students from the Industrial Engineering department is significantly larger than that of the Electrical Engineering department students, would you agree?

****

a) H_0: Major makes no difference when it comes to passing or failing EI exam

CORRECT APPROACH CALLS FOR GROUPING CELLS ($e_{ij} \geq 5$ FOR AT LEAST 95% OF THE CELLS)

DPMENT	PASS	FAIL	TOTALS
CE	13	12	25
IE	8	4	12
EE	37	10	47
ME	18	5	23
	76	31	107

THEREFORE, WE ARE ACTUALLY TESTING THE DEPARTMENTS, NOT THE MAJORS!

TEST STATISTIC

$$\chi^2 = \sum_i \sum_j \frac{(f_{ij} - e_{ij})^2}{e_{ij}} = n\left[\sum_i \sum_j \frac{f_{ij}^2}{n_{i.}\, n_{.j}} - 1\right]$$

$$= 6.45$$

$\chi^2 < \chi^2_{0.05,\,3} = 7.815$ CAN NOT REJECT H_0

NO SIGNIFICANT DIFFERENCES FOUND AMONG DEPARTMENTS

b) H_0: $p_{\text{failing IE}} = p_{\text{failing EE}}$ or $p_1 - p_2 = 0$

H_1: $p_{\text{failing IE}} > p_{\text{failing EE}}$ or $p_1 - p_2 > 0$

	pass	fail
IE	8	4
EE	37	10

$$\hat{p}_{\text{failing}} = \frac{4+10}{12+47} = \frac{14}{59}$$

$$S_{p_1-p_2} = \sqrt{\hat{p}(1-\hat{p})\left(\frac{1}{n_1}+\frac{1}{n_2}\right)}$$

$$= \sqrt{\frac{14}{59}\cdot\frac{45}{59}\left(\frac{1}{12}+\frac{1}{47}\right)} = 0.1376$$

$$\hat{p}_1 - \hat{p}_2 = \frac{4}{12} - \frac{10}{47} = 0.1206$$

$$p_1 - p_2 \leq (\hat{p}_1 - \hat{p}_2) \pm t_{0.05,\, n_1+n_2-2} \cdot S_{p_1-p_2}$$

$$\leq 0.1206 \pm (1.645)(0.1376)$$

$$\therefore \quad -0.106 < p_1 - p_2 < 0.347$$

Since zero is inside the confidence interval we can not reject H_0. We conclude that there is no significant difference between the proportions of student failing the IE exam among these departments (DO NOT AGREE WITH STATEMENT).

GOODNESS OF FIT TESTS

8-21

A statistics professor believes that the number of children in a family follow a Poisson distribution. He chooses 32 families at random and records the number of children in each. The data is as follows:

Number of children	0	1	2	3	4 or more
Number of families	4	10	8	7	3

Use the Chi-Square goodness of fit test to verify his assertion at the 95% confidence level.

****

The mean $= \lambda = \dfrac{0(5)+1(9)+2(8)+3(7)+4(3)}{32} = \dfrac{58}{32} = 1.8$

The Poisson Probability $P(x) = \dfrac{\lambda^x e^{-2}}{x!} = \dfrac{(1.8)^x e^{-1.8}}{x!}$

$P(x=0) = .1653 \quad P(x=1) = .2975 \quad P(x=2) = .2678 \quad P(x=3) = .1607$

$P(x \geq 4) = 1 - [P(0)+P(1)+P(2)+P(3)] = (1 - .8913) = .1087$

Number of Children (x)	P(x)	Expected σ_i 32P(x)	Observed o_i	$(o_i - \sigma_i)^2$
0	.1653	5.3	5	.017
1	.2975	9.5	9	.026
2	.2675	8.6	8	.042
3	.1607	5.1	7	.708
$\geq$ 4	.1087	3.5	3	.071

$\chi^2 = \Sigma\ .864$

$\chi^2_{Test} = .864 \quad \chi^2_{(4,.05)} = 9.48 \quad \chi^2_{Test} < \chi^2_{crit}$

Do not reject the hypothesis

8-22

The following data, taken from personnel records for 3 months, cover 730 man-days of lost work.

Day	Monday	Tuesday	Wednesday	Thursday	Friday
Days lost	175	140	138	126	151

How would you test whether the number of days lost is uniformly distributed throughout the week (e.g., that Mondays are not significantly different from, say, Thursdays, with respect to number of people absent?)

****

H_0 : Data comes from discrete uniform distribution,

e.g., $f_i = \frac{730}{5} = 146$

	Observed	Expected	$\frac{(o_i - e_i)^2}{e_i}$
Monday	175	146	5.760
Tuesday	140	146	0.247
Wednesday	138	146	0.438
Thursday	126	146	2.740
Friday	151	146	0.171
	730	730	9.356 $= \chi^2$

Test against $\chi^2_{5-1-1,\,0.05} = \chi^2_{3,0.05} = 7.815$

Since $\chi^2 > \chi^2_{3,0.05}$ reject H_0. ~~DATA~~ IS NOT UNIFORMLY DISTRIBUTED

8-23

The following table gives the grade distribution from a single professor in the same subject at consecutive hours. Test the hypothesis at the 5% level that there is no difference in grading.

	A	B	C	D	F	TOTAL
9:00	8	4	10	13	5	40
10:00	7	11	20	12	10	60

****

If there were no difference, we could compute expected values for each class:

	A	B	C	D	F
9:00	6	6	12	10	6
10:00	9	9	18	15	9

FOR EXAMPLE, the expected A's at 9:00 is obtained from the given table above by

$$\frac{(\text{ROW TOTAL})(\text{COLUMN TOTAL})}{\text{GRAND TOTAL}} = \frac{40 \cdot 15}{100} = 6$$

etc.

HERE $\chi^2 = \sum \frac{(X_{\text{EXPECTED}} - X_{\text{OBSERVED}})^2}{X_{\text{EXPECTED}}}$

$$= \frac{(6-8)^2}{6} + \frac{(9-7)^2}{9} \text{ etc} = 4.56$$

DEGREES OF FREEDOM ARE (NO. OF COLUMNS − 1)(NO. OF ROWS − 1)

$$= 4 \cdot 1 = 4$$

$\chi^2_{.05,4} = 9.49$, SINCE $4.56 < 9.49$, the differences in grading are not significant.

8-24

An inventory manager knows from his study of 100 randomly selected records that the lead time (time elapsed since an order is placed until the item is available on the shelf) for a certain provider has the following distribution:

Lead time (T)[days]	frequency
<20	46
$20 \leq T < 40$	19
$40 \leq T < 60$	17
$60 \leq T < 80$	12
$T \geq 80$	6
	100

Test whether the lead time may be regarded as an exponentially distributed random variable with a mean of 40 days. Use a five percent significance level.

****

Expected probabilities if $T \sim \exp(\beta = 40)$

$$Pr(T_1 < T < T_2) = \int_{T_1}^{T_2} \frac{1}{40} e^{-t/40}\,dt = e^{-T_1/40} - e^{-T_2/40}$$

T	probability $= p_i$	$e_i = 100 \cdot p_i$	$\frac{(o_i - e_i)^2}{e_i}$
< 20	$1 - e^{-0.5} = 0.393$	39.3	1.142
$20 \leq T < 40$	$e^{-0.5} - e^{-1} = 0.239$	23.9	1.005
$40 \leq T < 60$	$e^{-1} - e^{-1.5} = 0.145$	14.5	0.431
$60 \leq T < 80$	$e^{-1.5} - e^{-2} = 0.088$	8.8	1.164
≥ 80	$e^{-2} - 0 = 0.135$	13.5	4.167
	1.000	100.0	7.908

Test against $\chi^2_{4,0.05} = 9.488$

$$\chi^2 = \sum_i \frac{(o_i - e_i)^2}{e_i} < \chi^2_{4,0.05}$$

CAN NOT REJECT HYPOTHESIS

8-25

PERFORM A GOOD OF FIT TEST WITH THE TRIANGULAR DISTRIBUTION FOR THE FOLLOWING DATA :

VALUE	0	1	2	3	4	5	6	7	8	9
COUNT	0	0	4	12	18	25	20	14	7	0

FOR THE TRIANGULAR DISTRIBUTION :

P(X) = 0.0625 (X - 1)   FOR 1 < X <= 5
= 0.0625 (9 - X)   FOR 5 <= X < 9

****

THE FIRST STEP IS TO CALCULATE THE THEORETICAL FREQUENCY COUNT FOR THE DISTRIBUTION. USING THE FUNCTIONS FOR P(X), FIRST CALCULATE THE PROBABILITY OF A X VALUE THEN MULTIPLY BY 100, THE TOTAL COUNT OF OBSERVATIONS. DOING THIS, THE RESULT IS :

VALUE	0	1	2	3	4	5	6	7	8	9
PROB	0	0	6.25	12.5	18.75	25	18.75	12.5	6.5	0

NEXT, THE CHI^2 STATISTIC IS CALCULATED AND COMPARED TO THE TABLED VALUE.

(NOTE FOR ANY VALUE THAT HAS LESS THAN 5 OBSERVATIONS, IT IS COMBINED WITH AN ADJACENT VALUE.)

$$\chi^2 = \frac{(4+12-6.25+12.5)^2}{18.75} + \frac{(18-18.75)^2}{18.75} + \frac{(25-25)^2}{25} + \frac{(20-18.75)^2}{18.75} + \frac{(14-12.5)^2}{12.5} + \frac{(7-6.25)^2}{6.25} = 0.786$$

$$\nu = 6-5 = 5 \qquad \alpha = 0.05 \qquad \chi^2_{5,.05} = 11.075$$

SINCE $\chi^2 < \chi^2_{5,.05} \longrightarrow$ GOOD FIT !

8-26

During the study of crossing of two different species of flower plants, 100 hybrid flower plants were observed with three different flower colors having frequencies of 22%, 51%, and 27%, respectively. If the corresponding theoretical frequencies are, 30%, 40%, and 30%, respectively, determine if there is a significant difference between the observed and theoretical frequencies.

****

This is a case of a Chi-square distribution.

The Null Hypothesis H_0: There is no significant difference between the observed and theoretical frequencies.

The Alternate Hypothesis H_1: There is a difference between the observed and theoretical frequencies.

Let O = observed frequency and E = expected or theoretical frequency. The computations may be tabulated as follows.

O	E	O − E	$(O-E)^2$	$(O-E)^2/E$
22	30	−8	64	2.1
51	40	11	121	3.0
27	30	−3	9	0.3

$$\chi^2 = \sum \frac{(O-E)^2}{E} = 5.4$$

Degrees of freedom = d.f. = $n-1 = 3-1 = 2$

From the Chi-Square distribution tables for $\chi^2 = 5.4$, with 2 d.f., the probability is given as:

$$0.05 < p < 0.10$$

At the 5% level H_0 is acceptable. This means there is no significant difference between the observed and theoretical frequencies.

8-27

A quality control engineer takes daily samples of 10 electronic components, checking them for imperfections. If on 200 consecutive working days he obtained 112 samples with 0 defectives, 76 with 1 defective and 12 with 2 defectives, test at the .05 level whether these samples may be looked upon as samples from a binomial distribution.

****

TOTAL NO. OF DEFECTIVES = $112(0) + 76(1) + 12(2) = 100$

AND $\hat{p} = \dfrac{100}{10(200)} = 0.05$

FROM BINOMIAL TABLES, for a p of .05, $n = 10$:

$P(0) = .5987$, $P(1) = .3152$, $P(2 \text{ OR MORE}) = .0861$

therefore

f_{obs}	$f_{expected}$
112	119.7
76	63.0
12	17.2

and $\chi^2 = \sum \frac{(f_{OBS} - f_{EXP})^2}{f_{EXP}}$

$= .50 + 2.68 + 1.57$

$= 4.75$

But $\chi^2_{.05, 1 d.f.} = 3.84 < 4.75$

∴ We conclude the sample did not come from a binomial distrubution

PROPORTIONS

8-28

In an experiment in botony the results of crossing two hybrids of a species of flower gave observed frequencies of 123, 50, 36 and 15. Do these results disagree with theoretical frequencies which specify a 9:3:3:1 ratio? Use α = .05.

****

TOTAL FREQUENCY = 224

EXPECTED FREQUENCIES ARE $\frac{9}{16}N, \frac{3}{16}N, \frac{3}{16}N, \frac{1}{16}N$

OR 126, 42, 42, 14

$$\chi^2 = \sum \frac{(FREQ_{EXP} - FREQ_{OBS})^2}{FREQ_{EXP}} = \frac{3^2}{126} + \frac{8^2}{42} + \frac{6^2}{42} + \frac{1^2}{14}$$

$= 2.52$. BUT $\chi^2_{.05,3} = 7.81$; AND THEREFORE the results do not disagree with theory.

9 VARIANCES

POINT ESTIMATES

9-1

A BALL BEARING MUST HAVE A TOLERANCE OF +- 0.0006 INCHES. A SAMPLE OF 10 BALL BEARINGS ARE MEASURED AND THE SAMPLE VARIANCE WAS 0.00000005 IN.^{2}

AT AN ALPHA LEVEL OF 0.05, CAN WE SAY THAT THE MANUFACTURING PROCESS IS UNDER CONTROL ?

****

$\alpha = 0.05 \quad \eta = 10 - 1 = 9 \quad S^2 = 0.00000005 \text{ in}^2$

$S = 0.000224 \text{ in}$

TOLERANCE $= \pm 0.0006 \longrightarrow$ BETWEEN $\pm 2\sigma$ AND $\pm 3\sigma$

$\therefore \sigma = 0.002$ OR 0.003

$$\chi^2 = \frac{(N-1)S^2}{\sigma^2} = \frac{(9)(.00000005)}{(.0002)^2} = 11.25$$

$$= \frac{(9)(.00000005)}{(.0003)^2} = 5$$

$$\chi^2_{9,.05} = 16.919$$

$\therefore$ PROCESS IS UNDER CONTROL WITH EITHER σ

9-2

If S_1^2 and S_2^2 represent the variances of independent random samples of size $n_1 = 8$ and $n_2 = 12$, taken from Normal populations with equal variances, find $Pr(S_1^2/S_2^2 < 4.89)$.

****

$X_1 \sim N(\mu_1, \sigma)$ Sample Size $n_1 = 8$

$X_2 \sim N(\mu_2, \sigma)$ Sample Size $n_2 = 12$

$Pr\left[\frac{S_1^2}{S_2^2} < 4.89\right] = Pr(F < 4.89)$ with $\nu_1 = 7$ & $\nu_2 = 11$

$= 0.99$ from F distribution table (non-shaded area under p.d.f.)

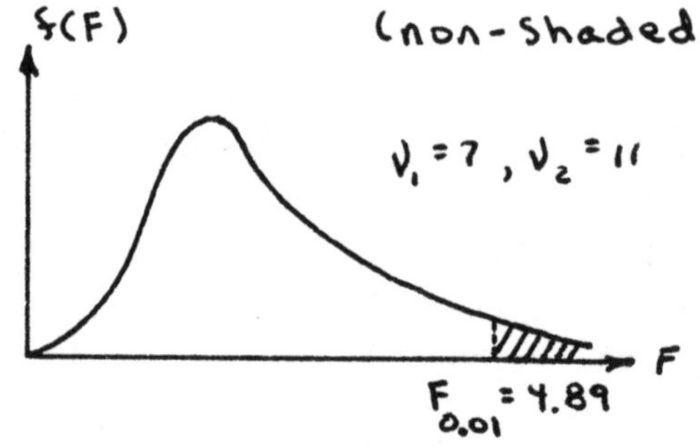

9-3

Find the probability that a random sample of 25 observations from a Normal population with variance of 6 will have a sample variance greater than 9.1.

****

The random variable $\frac{(n-1)s^2}{\sigma^2}$ is Chi-Square distributed with $n-1$ d.f.

s^2 - Sample Variance

σ^2 - Population Variance

$Pr(s^2 > 9.1) = Pr\left[\frac{(n-1)s^2}{\sigma^2} > \frac{24(9.1)}{6}\right]$

$= Pr(\chi^2 > 36.4)$ $\nu = n-1 = 24$ d.f.

$= 0.05$ from Chi-square table

9-4

The following values represent parts per million of the dreaded virus Betchakillus measured in a sample of several rural well water sources:

15, 18, 12, 12, 10, 14, 16, 11, 18, 12, 8, 10

Estimate the standard deviation of the data by using the mean range of subgroups of size 3, and compare this estimate to the calculated standard deviation estimate for the population using an appropriate formula.

****

Range values of subgroups of size 3:

$$\underbrace{15 \quad 18 \quad 12}_{6} \quad \underbrace{12 \quad 10 \quad 14}_{4} \quad \underbrace{16 \quad 11 \quad 18}_{7} \quad \underbrace{12 \quad 8 \quad 10}_{4}$$

Population standard deviation estimate is denoted by s, where

$$s \cong \bar{R} \cdot d$$

($\bar{R}$ is average range of subgroup range values, d is value of "range coefficient" available from table for given subgroup size*)

$$s \cong \left(\frac{6+4+7+4}{4}\right)(0.5908)$$

(d from table in textbook for subgroup size $n = 3$)

$$s \cong (5.25)(0.5908)$$

$$\underline{\underline{s \cong 3.10}}$$

Using the 12 data values to find s,

$$s = \sqrt{\frac{(12)\Sigma x^2 - (\Sigma x)^2}{(12)(11)}}$$

$\Sigma x^2 = 2{,}142 \qquad \Sigma x = 156$

$$s = \sqrt{\frac{(12)(2{,}142) - (156)^2}{(12)(11)}}$$

$$\underline{\underline{s = 3.22}} \quad \text{(compare to 3.10 above)}$$

* Note that $s \cong \bar{R} \cdot d$ in some texts appears as $s \cong \bar{R}/d_2$, where d_2 is obtained from a table. Also note that d is close to the value $1/\sqrt{n}$ for small subgroups of size n.

9-5

A sample of 400 electric fuses was drawn from a large batch and each one was tested. It was found that 40 fuses were defective. Determine a point estimate for the proportion of defective fuses in the batch, and a point estimate for the standard deviation of the defective proportion in a random sample of 400 fuses.

****

A point estimator $\bar{p}$ for the proportion of defective fuses in the batch = proportion of defective fuses in one sample. Hence

$$\bar{p} = 0.1$$

Suppose N = number of bad fuses in a sample of m

This random variable obeys the binomial probability distribution;

$$P(N=n) = p(n) = {}^{m}C_{n}\, p^{n}(1-p)^{m-n} \qquad n = 0, 1, \cdots m$$

where

p = probability of picking a bad fuse from the batch.

It is known that

Mean $E(N) = mp$ and $Var(N) = mp(1-p)$

Hence

$$\mu = E\left(\frac{N}{m}\right) = \frac{E(N)}{m} = \frac{mp}{m} = p$$

$$\sigma^2 = Var\left(\frac{N}{m}\right) = \frac{Var(N)}{m^2} = \frac{mp(1-p)}{m^2} = \frac{p(1-p)}{m}$$

Hence a point estimator for σ^2 is

$$\bar{\sigma}^2 = \frac{\bar{p}(1-\bar{p})}{m}$$

For the given data $\bar{\sigma} = \sqrt{\dfrac{0.1(1-0.1)}{400}} = 0.015$

9-6

A random sample of five observations from a normal population yielded a minimum value of 30 and a maximum value of 35. Estimate the population variance.

****

$$\hat{\sigma}^2 = \left(\frac{R}{d_2}\right)^2 = \left(\frac{5}{2.326}\right)^2 = 4.622$$

this point estimate is usually used in statistical quality control applications.

INTERVAL ESTIMATES

9-7

A consumer group wishes to investigate the consistency of gasoline consumption efficiency (measured in miles per gallon) for a particular make of automobile during city driving. Ten trials with the automobile yielded the following results in miles per gallon.

23.2, 25.3, 22.9, 23.7, 24.6, 23.0, 24.2, 22.5, 23.6, 24.4

Assuming that the automobile's city driving mileage is normally distributed, construct a 95% confidence interval for the standard deviation of miles per gallon for all cars of this type.

****

$n = 10 \qquad s^2 = 0.769$

$$\frac{(n-1)s^2}{\chi_2^2} < \sigma^2 < \frac{(n-1)s^2}{\chi_1^2}$$

$$\frac{(9)(.769)}{19.023} < \sigma^2 < \frac{(9)(.769)}{2.700} \qquad 0.364 < \sigma^2 < 2.563$$

$$\therefore \quad 0.603 < \sigma < 1.601$$

9-8

From 300 experiments the average was computed to be 28, and the standard deviation was computed to be 5. Find the two-sided 95% confidence interval for the standard deviation.

****

$n = 300$, $\bar{X} = 28$, $\alpha = 0.05$
$df = 299$, $s = 5$

$\chi^2 = (n-1)s^2/\sigma^2$ IS A CHI-SQUARED RANDOM VARIABLE. SINCE n IS SO LARGE, CONSIDER $\sqrt{2\chi^2}$ WHICH IS A NORMAL RANDOM VARIABLE WITH $\mu = \sqrt{2(df)-1}$ AND $\sigma^2 = 1$.

$$Pr\{\chi_L^2 \le \chi^2 \le \chi_u^2\} = Pr\{\sqrt{2\chi_L^2} \le \sqrt{2\chi^2} \le \sqrt{2\chi_u^2}\}$$

$$= Pr\{\sqrt{2\chi_L^2} - \sqrt{597} \le \sqrt{2\chi^2} - \sqrt{597} \le \sqrt{2\chi_u^2} - \sqrt{597}\}$$

$$= Pr\{-z_{.025} \le Z \le z_{.025}\} = .95$$

$$\sqrt{2\chi_L^2} - \sqrt{597} = -1.96 \quad \& \quad \sqrt{2\chi_u^2} - \sqrt{597} = 1.96$$

$$\chi_L^2 = 252.5 \quad , \quad \chi_u^2 = 348.3$$

$$252.5 \le \frac{(n-1)s^2}{\sigma^2} \le 348.3$$

$$4.63 \le \sigma \le 5.44$$

9-9

A random sample from a Normal population resulted in the following values:

$X_1 = 10 \quad X_2 = 9 \quad X_3 = 13 \quad X_4 = 8 \quad X_5 = 10$

Find the 95% confidence interval for the population variance σ^2.

****

The $100(1-\alpha)\%$ confidence interval for σ^2 is obtained as follows:

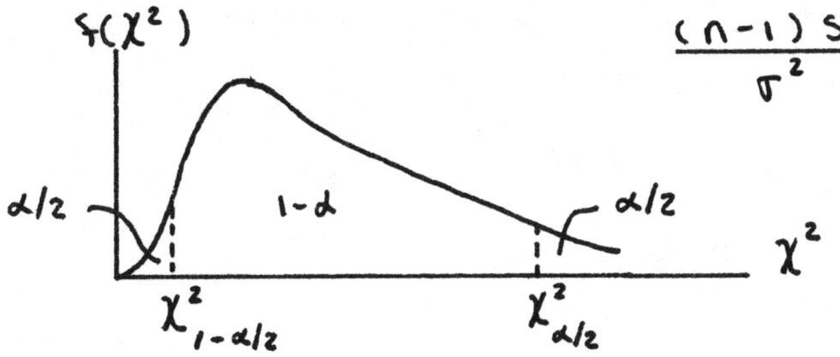

$\frac{(n-1)s^2}{\sigma^2}$ is χ^2 distributed with $n-1$ d.f.

σ^2 - Normal population var.
s^2 - Sample variance

$$Pr\left[\chi^2_{1-\alpha/2} \le \frac{(n-1)s^2}{\sigma^2} \le \chi^2_{\alpha/2}\right] = 1-\alpha$$

from which it follows, $Pr\left[\frac{(n-1)s^2}{\chi^2_{1-\alpha/2}} \geq \sigma^2 \geq \frac{(n-1)s^2}{\chi^2_{\alpha/2}}\right] = 1-\alpha$

Hence, the $100(1-\alpha)\%$ C.I. for σ^2 is $\left[\frac{(n-1)s^2}{\chi^2_{\alpha/2}}, \frac{(n-1)s^2}{\chi^2_{1-\alpha/2}}\right]$

From the data, $\bar{x} = 10$ & $s^2 = 3.5$

$\alpha = 0.05$, 95% LCL $= \frac{4(3.5)}{\chi^2_{0.025}} = \frac{4(3.5)}{11.143} = 1.26$

95% UCL $= \frac{4(3.5)}{\chi^2_{0.975}} = \frac{4(3.5)}{0.484} = 28.93$

where $\chi^2_{0.025}$ and $\chi^2_{0.975}$ are obtained from Chi-Square table using $\nu = n-1 = 4$ d.f.

HYPOTHESES TESTS: ONE VARIANCE

9-10

A sample from an assumed normal distribution gave the set of values: 9, 14, 10, 12, 7, 13, 11, 12. a) What is the single best unbiased estimate of μ? of σ^2? b) Find an 80% confidence interval for μ. c) Test H_o: $\mu = 10$ against the alternative $\mu \neq 10$ at the 5% level of significance, if $\sigma = 2$. d) Test H_o: $\mu = 13$ against $\mu < 13$ at the 5% level. e) Test H_o: $\sigma^2 = 2$ against $\sigma^2 > 2$ at the 5% level.

****

(a) $\mu_{EST} = \bar{X} = 11 \qquad \sigma^2_{EST} = s^2 = 36/7 = 5.14$

(b) $11 \pm t_{.10,7} \cdot \dfrac{s}{\sqrt{n}} = 11 \pm \dfrac{(1.415)(3)}{3.742} = 11 \pm 1.134$

(c) $Z = \dfrac{1}{2/\sqrt{8}} = \sqrt{2} = 1.414 < 1.96$ OR $Z_{.025}$

CANNOT REJECT Ho

(d) $t = \dfrac{-2\sqrt{8}}{\sqrt{5.14}} = -2.495 < -1.895$ OR $-t_{.05,7}$

REJECT Ho

(e) $\chi^2 = \dfrac{(n-1)s^2}{\sigma^2} = \dfrac{7(5.14)}{2} = 18 > 14.067$ OR $\chi^2_{.05,7}$

REJECT Ho

9-11

Modern Welding Company is testing the shear strength of a particular weld. They would like the strength of such welds to have little variability. Does the sample of weld strengths given below show sufficient evidence to allow Modern Welding to claim that the variance of shear strength is no more than 950 at the 0.05 level of significance?

SHEAR STRENGTH (in pounds): 2190, 2280, 2285, 2275, 2340, 2305, 2250
2235, 2270, 2280

****

$n = 10$, $\bar{X} = 2271$, $\alpha = 0.05$
$df = 9$, $s = 40.26$

$H_0: \sigma^2 = 950$
$H_A: \sigma^2 > 950$

THE TEST STATISTIC IS A CHI-SQUARED, ONE-SIDED.

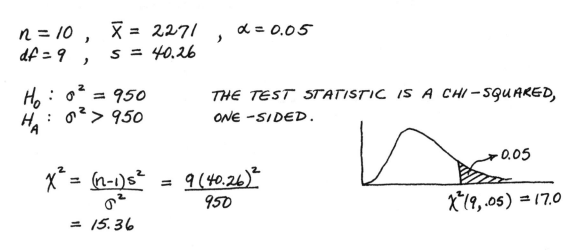

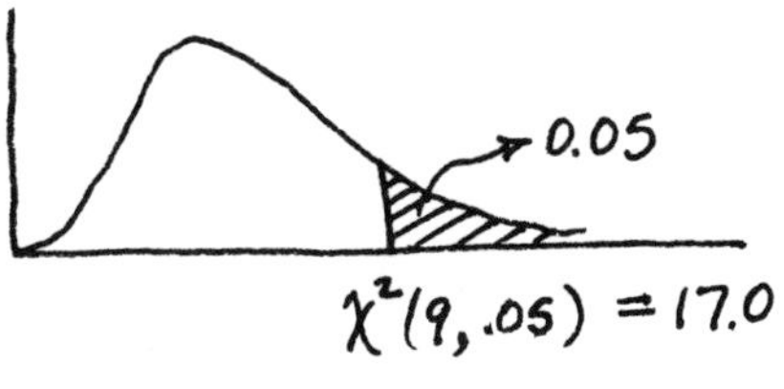

$$\chi^2 = \frac{(n-1)s^2}{\sigma^2} = \frac{9(40.26)^2}{950}$$
$$= 15.36$$

SINCE $15.36 < 17.0$, FAIL TO REJECT H_0.

9-12

Given that x is a normal variable and given the sample values $\bar{x} = 42$, $S = 6$, $n = 20$: a) Test H_o: $\sigma = 9$, H_1: $\sigma < 9$ at $\alpha = .05$. b) Test H_o: $\mu = 44$, H_1: $\mu \neq 44$ at $\alpha = .05$.

****

a) H_o: $\sigma = 9$, $H_1: \sigma < 9$ with $n = 20$, $\alpha = .05$

TEST STATISTIC : $\chi^2 = \frac{(n-1)s^2}{\sigma^2} = \frac{19 s^2}{81}$

CRITICAL REGION: $\chi^2 < \chi^2_{.95, 19} = 10.117$

COMPUTED $\chi^2 = \frac{19 \cdot 36}{81} = 8.44 < 10.117$ Reject Ho

b) Ho: $\mu = 44$, $H_1: \mu \neq 44$ with $n = 20$, $\alpha = .05$

TEST STATISTIC : $t = \frac{\bar{x} - \mu_0}{s/\sqrt{n}} = \frac{(\bar{x} - 44)\sqrt{20}}{s}$

CRITICAL REGION : $|t| > t_{.025, 19} = 2.093$

COMPUTED $t = \frac{2\sqrt{20}}{6} = 1.49 < 2.093$

therefore we cannot reject Ho.

9-13

An optics manufacturer supplies a U. S. Navy contractor with prisms for submarine periscopes. One of the prism manufacturing quality standards is that the standard deviation of the prism base diameter must be less than 0.003 inches. Placing the burden of proof on the prism manufacturer, test at the 1% level of significance if prisms are meeting the quality standard if a random sample of 20 prisms yields a diameter standard deviation of 0.00173 inches.

****

$H_0: \sigma = 0.003$ Test Statistic: χ^2

$H_1: \sigma < 0.003$ $\chi^2_{.99,19} = 7.633$

Reject H_0 if $\chi^2_{cal} < \chi^2_{.99,19}$

$$\chi^2_{cal} = \frac{(19)(0.00173)^2}{(0.003)^2} = 6.32 < 7.633$$

Reject H_0 and conclude that prisms are meeting the quality standard.

9-14

From past experience, the variance in diameters of shafts was 0.0052. A new machinist turns out 100 shafts with $S^2 = .0065$. Does this indicate greater variability than normal for this man's work?

****

$H_0: \sigma^2 = .0052$ $H_1: \sigma^2 > .0052$

$$\chi^2 = 100\left(\frac{.0065}{.0052}\right) = 125$$

USING $\sqrt{2\chi^2} - \sqrt{2m-1} \sim$ Normal with $\sigma = 1$

$$\sqrt{250} - \sqrt{199} = 15.81 - 14.10 = 1.71 > Z_{.05} = 1.645$$

∴ We can reject Ho, accept H_1 and say that this indicates greater variability

9-15

The variance of a Normal population was 80 before certain changes were made. A random sample of 26 observations yielded a sample variance of 100. Did the changes increase the variance?

****

Let σ^2 = Variance after changes

$H_0: \sigma^2 = \sigma_0^2 = 80$

$H_1: \sigma^2 > \sigma_0^2$

The test statistic is $\chi^2 = \dfrac{(n-1)s^2}{\sigma^2}$

Under the null hypothesis H_0, i.e. H_0 is true

$Pr(\chi^2 > \chi^2_{CR} \mid H_0 \text{ True}) = \alpha$, where α is the risk or probability of a Type I error

Since there is no risk specified, we shall choose a value of $\alpha = 0.05$

The critical value χ^2_{CR} is determined from the Chi-square distribution.

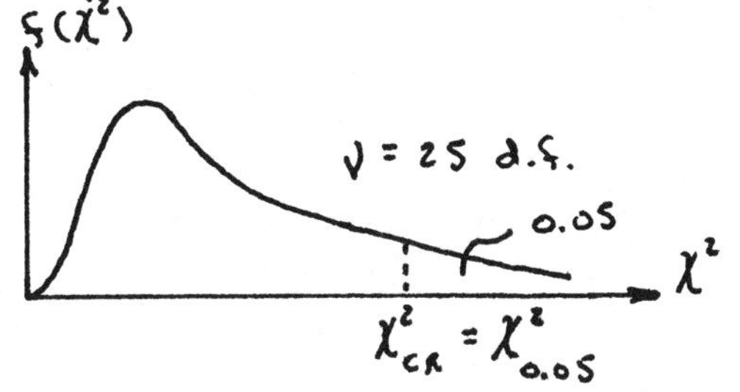

$\chi^2_{CR} = 37.652$

The decision rule is: Reject H_0 if $\dfrac{(n-1)s^2}{\sigma_0^2} > 37.652$

Equivalently, Reject H_0 if $s^2 > \dfrac{37.652\,\sigma_0^2}{n-1} = 120.5$

Since $s^2 = 100$ which is less than 120.5, we cannot reject H_0, i.e. the sample data does not support the conclusion that the population variance has increased.

HYPOTHESES TESTS: TWO VARIANCES

9-16

Two assembly line trainees are evaluated for assembly speed by performing a certain benchmark test. Ten trials by worker X yielded a mean processing time of 3.2 minutes with a standard deviation of 0.3 minutes. Worker Y performed the test 13 times and attained a mean time of 3.8 minutes with standard deviation of 0.4 minutes. From a previous experiment it is known that the processing time for the task is normally distributed for all trainees, although it is not known if the standard deviation of processing time is constant for all trainees. Using a 10% level of significance, test the hypothesis that there is no significant difference in the mean processing times of the two workers. Note: It is appropriate to test the hypothesis that the two population variances are equal before the two-sample t test is used for means.

****

$H_0: \sigma_X^2 = \sigma_Y^2$

$H_1: \sigma_X^2 \neq \sigma_Y^2$

$$F_{cal} = \frac{s_Y^2}{s_X^2} = \left(\frac{0.4}{0.3}\right)^2 = 1.78$$

$$F_{.05,12,9} = 3.07$$

Since $F_{cal} < F_{\alpha/2}$, cannot reject H_0 and conclude that the two-sample t test is appropriate.

$H_0: \mu_X = \mu_Y$

$H_1: \mu_X \neq \mu_Y$

$$t_{cal} = \frac{-0.6}{\sqrt{(9)(0.3)^2 + (12)(0.4)^2}} \sqrt{\frac{(130)(21)}{23}}$$

$$t_{cal} = -3.96$$

Since $t_{cal} < -t_{\alpha/2} = -t_{.05,21} = -1.721$, reject H_0 and conclude that there is a significant difference in the means.

9-17

The mean and standard deviation in a blood pressure sample of 11 patients were 145 and 14.0 [mm Hg], respectively. A second group of 13 patients is given a certain drug and they showed $\bar{x}_2 = 140$ with $s_2 = 9.0$. Is the drug effective in lowering blood pressure?

****

$n_1 = 11 \qquad \bar{x}_1 = 145 \qquad s_1 = 14$

$n_2 = 13 \qquad \bar{x}_2 = 140 \qquad s_2 = 9$

FIRST TEST $\quad \sigma_1^2 = \sigma_2^2$ (unknown but equal)

$$\frac{s_1^2/s_2^2}{F_{0.05,\, n_1-1,\, n_2-1}} < \frac{\sigma_1^2}{\sigma_2^2} < \frac{s_1^2/s_2^2}{F_{0.95,\, n_1-1,\, n_2-1}}$$

$$\frac{2.42}{2.75} < \frac{\sigma_1^2}{\sigma_2^2} < \frac{2.42}{1/2.91}$$

$$0.88 < \frac{\sigma_1^2}{\sigma_2^2} < 7.04$$

$\sigma_1^2 = \sigma_2^2 \Rightarrow \frac{\sigma_1^2}{\sigma_2^2} = 1$: inside interval, accept $\sigma_1^2 = \sigma_2^2$

$\bar{x}_1 - \bar{x}_2 = 5 \qquad s_{\bar{x}_1 - \bar{x}_2} = \sqrt{\frac{(n_1-1)s_1^2 + (n_2-1)s_2^2}{n_1+n_2-2}\left(\frac{1}{n_1}+\frac{1}{n_2}\right)}$

$$= 4.729$$

$$\mu_1 - \mu_2 < (\bar{x}_1 - \bar{x}_2) \pm t_{n_1+n_2-2} \cdot s_{\bar{x}_1-\bar{x}_2}$$

$-4.8 < \mu_1 - \mu_2 < 14.8 \qquad$ since $t_{0.025,22} = 2.074$

since zero is inside interval, the hypothesis $\mu_1 - \mu_2 = 0$ can not be rejected. DRUG IS NOT EFFECTIVE.

9-18

The Benkelman beam method is widely used for determining the deflection of existing highway pavement sections. In a study of a State Highway Department, Benkelman beam tests have been conducted on two sections of a pavement, resulting in the following sample information. For the first sample, 10 observations of the rebound deflection were taken, yielding a mean of 26×10^{-3} in. and a standard deviation of 2×10^{-3} in. For the second section, 14 measurements were taken, yielding a mean of 21×10^{-3} in. and standard deviation of 7×10^{-3} in.

You may assume that deflection measurements for each section follow a normal distribution.

a) Based on the above sample information, can you conclude with 98% confidence that the respective (population) variances of deflection measurements for the two pavement sections are different?

b) Using a significance level of 0.01, determine whether the two sections have the same mean rebound deflections. Make sure that the test statistic used and its distribution are consistent with your answer to part (a).

****

a) $H_0: \sigma_1^2 = \sigma_2^2$ vs. $H_1: \sigma_1^2 \neq \sigma_2^2$; TWO-SIDED TEST @ $\alpha = 0.02$.

σ_1^2, σ_2^2 denote the population variances of populations 1 and 2 respectively
s_1^2, s_2^2 " " sample variances
n_1, n_2 " " respective sample sizes.

Test statistic: $\frac{s_1^2}{s_2^2} \sim F(n_1 - 1, n_2 - 1)$

Critical Region: Reject H_0 if $\frac{s_1^2}{s_2^2} < F_{9,13,0.99}$ or $\frac{s_1^2}{s_2^2} > F_{9,13,0.01}$

i.e. if $\frac{s_1^2}{s_2^2} < 1/5.06$ or $\frac{s_1^2}{s_2^2} > 4.19$

For given data,

$$F^* = \frac{s_1^2}{s_2^2} = \frac{4}{49} = 0.08 < \frac{1}{5.06}$$ ⇒ REJECT H_0; variances are different.

b) $H_0: \mu_1 - \mu_2 = 0$ vs. $H_1: \mu_1 - \mu_2 \neq 0$.

test statistic,
given that populations are normal, sample sizes are small and underlying variances are different,

$$T = \frac{\bar{X}_1 - \bar{X}_2}{\sqrt{\frac{S_1^2}{n_1} + \frac{S_2^2}{n_2}}} \sim t(\nu)$$

under H_0,
test statistic is t-distributed, with v degrees of freedom,
with

$$v = \frac{\left(\frac{s_1^2}{n_1} + \frac{s_2^2}{n_2}\right)^2}{\frac{(s_1^2/n_1)^2}{n_1+1} + \frac{(s_2^2/n_2)^2}{n_2+1}} - 2$$

For the given values of s_1, s_2, n_1 and n_2,

we obtain $v = 16.3$; use $v = 17$

Critical region: reject H_0 if

$$T^* < -t_{0.005,17} \text{ or } T^* > t_{0.005,17}$$

i.e. if $T^* < -2.898$ or $T^* > 2.898$

For given data,

$$T^* = \frac{26\times 10^{-3} - 21\times 10^{-3}}{\sqrt{\frac{4\times 10^{-6}}{10} + \frac{49\times 10^{-6}}{14}}} = 2.532$$

Since $-2.898 < 2.532 < 2.898$,

DO NOT REJECT
the null hypothesis that
the two sections have the
same mean rebound deflections.

9-19

The flow rate in a small stream (in cubic feet per minute) was measured on two days with the following results, several measurements being taken on each day:

Day 1: 750, 720, 725, 760, 755, 730, 740
Day 2: 735, 728, 723, 722, 728, 733, 725, 733, 725

Determine if there is significant difference at the 5% level in flow rate variances on the two days the observations were taken.

****

The F test may be used to compare variance estimates s_1^2 and s_2^2 on Days 1 and 2 as follows:

$$s_1^2 = \frac{(7)(\Sigma \text{ rates}^2) - (\Sigma \text{ rates})^2}{(7)(6)} = \frac{(7)(3{,}834{,}650) - (5{,}180)^2}{(7)(6)} = 241.67$$

$$s_2^2 = \frac{(9)(4{,}770{,}034) - (6{,}552)^2}{(9)(8)} = 22.25$$

$$F = \frac{s_1^2}{s_2^2} = \frac{241.67}{22.25} = 10.861 > 3.58$$ (Table F value at 5% level with degrees of freedom 6 for numerator, 8 for denominator)

Conclusion: Significant difference in variances.

(It is interesting to note that a modified t-test will show no significant difference in mean flow rates on the two days.)

10 REGRESSION

SIMPLE LINEAR REGRESSION

10-1

The resistance of a new resistor was measured at various temperatures as shown below.

a) Find the regression of the resistance on temperature.
b) Estimate the variance of the resistance.
c) What value of resistance should be estimated when T = 24°?

TEMPERATURE (^{o}C)	10.0	15.0	20.0	25.0	30.0	35.0
RESISTANCE (ohms)	38.6	40.1	41.9	44.8	45.6	47.7

****

THE TEMPERATURE T IS THE INDEPENDENT VARIABLE, AND THE RESISTANCE R IS THE DEPENDENT VARIABLE. A LINEAR REGRESSION YIELDS

$$\hat{R} = bT + a .$$

$$b = \frac{\Sigma(x_i-\bar{x})(Y_i-\bar{Y})}{\Sigma(x_i-\bar{x})^2} = 0.371 , \quad a = \bar{Y} - b\bar{x} = 34.772$$

$$\bar{x} = 22.5 , \quad \bar{Y} = 43.12$$

a) $\hat{R} = 0.371T + 34.772$

ERROR SUM OF SQUARES $= e = \Sigma(Y_i - \hat{Y})^2 = 1.42$

b) $\sigma_Y^2 \sim S_Y^2 = \left(\frac{e}{n-2}\right) = 0.355$

c) $\hat{R} = 0.371(24) + 34.772 = 43.7$

10-2

ONE OF THE CROSS-SECTIONS OF A RIVER UNDER STUDY YIELDED THE FOLLOWING VALUES OF VELOCITY VERSUS WATER DEPTH.

WHAT IS THE BASIC RELATIONSHIP BETWEEN THE TWO VARIABLES AND WHAT ARE THE ESTIMATED PARAMETERS ?

VEL M/SEC	Q M^3/SEC	VEL M/SEC	Q M^3/SEC
0.10	4.8	0.215	88
0.14	12.5	0.24	112
0.18	23.5	0.265	165
0.21	37.0	0.280	210
0.24	52.0	0.300	255
0.27	70.0	0.320	308

****

THE FIRST IMPULSE MAY BE TO FORM A SIMPLE LINEAR RELATIONSHIP. IF SO THE RESULT IS :

$$VEL = 0.169 + 0.0054\ Q \qquad r = 0.718$$

HOWEVER, THE FIRST STEP, ESPECIALLY WITH UNKNOWN DATA SHOULD BE TO PLOT IT ! THE RESULT IS:

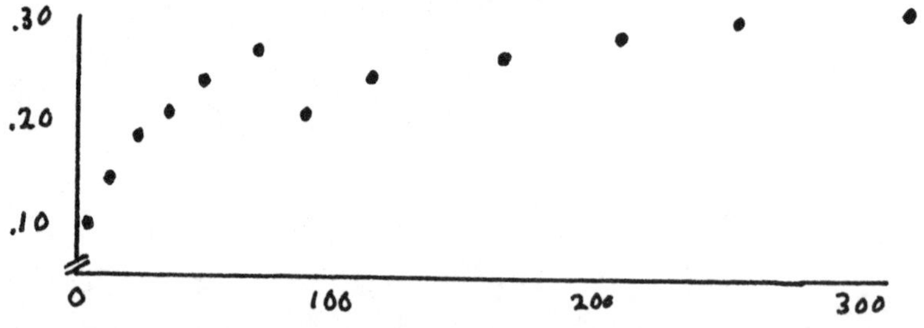

THE FUNCTION IS NOT ONLY NONLINEAR BUT HAS TWO PARTS TO IT. (WHERE THE RIVER LEAVES ITS NORMAL CHANNEL AND ENTERS THE FLOOD BANKS.) BY REALIZING THAT THE FUNCTION SHOULD BE OF THE FORM:

$$VELOCITY = A \times QTY^B$$

WE CAN TAKE LOG TO FORM A LINEAR RELATIONSHIP:

$$LN(VEL) = LN(A) + B \times LN(QTY)$$

NOW THE RESULTS ARE :

$$V = 0.0555\ Q^{.37}, \quad Q \leq 80 \qquad V = 0.0567\ Q^{.3}, \quad 80 < Q$$

$$r = .989 \qquad\qquad r = .993$$

10-3

Fit the equation of a straight line to the following data so that y values may be predicted for given values of x:

x	y
2.5	4.32
3.0	4.83
3.5	5.27
4.0	5.74
4.5	6.26

****

Let $y = a + bx$, where b is the slope and a the intercept. For linear regression, with 5 values,

$$a = \frac{\Sigma y \, \Sigma x^2 - \Sigma x \, \Sigma xy}{5\, \Sigma x^2 - (\Sigma x)^2} \qquad b = \frac{5\, \Sigma xy - \Sigma x \, \Sigma y}{5\, \Sigma x^2 - (\Sigma x)^2}$$

x	y	x^2	$x \cdot y$
2.5	4.32	6.25	10.800
3.0	4.83	9.00	14.490
3.5	5.27	12.25	18.445
4.0	5.74	16.00	22.960
4.5	6.26	20.25	28.170
17.5	26.42	63.75	94.865

$$a = \frac{(26.42)(63.75) - (17.5)(94.865)}{(5)(63.75) - (17.5)^2} = 1.931$$

$$b = \frac{(5)(94.865) - (17.5)(26.42)}{(5)(63.75) - (17.5)^2} = 0.958$$

Thus, $y = 1.931 + 0.958\,x$

10-4

An engineering professor noted that his three children of ages 16, 14, and 8 had heights of 69, 62, and 48 inches, respectively. Based on these values:

a) Develop a linear regression equation to predict height of a child as a fraction of age.
b) Use the results of a) to predict the height of an 11 year old child.
c) Use the results of a) to predict the height of a 30 year old person.
d) Comment on the results of b) and c).

****

a) $\Sigma x = 38 \quad \Sigma x^2 = 516 \quad \Sigma y = 179 \quad \Sigma y^2 = 10{,}909$

$\bar{y} = 59.67 \quad \bar{x} = 12.67 \quad \Sigma xy = 2356$

$$S_{xx} = 3(516) - (38)^2 = 104$$

$$S_{xy} = 3(2356) - (38)(179) = 266$$

$$b = 266/104 = 2.56$$

$$a = 59.67 - (2.56)(12.67) = 27.23$$

AND EQUATION IS

$$\text{HEIGHT} = 27.23 + 2.56 * \text{AGE}$$

b) HEIGHT AT 11 = 27.23 + 2.56(11) = 55.4

c) HEIGHT AT 30 = 27.23 + 2.56(30)

= 104.0 = 8.67 FEET

d) THE RESULTS SHOW THE DANGER OF APPLYING A REGRESSION EQUATION TO VALUES OUTSIDE THE ORIGINAL LIMITS OF THE DATA.

10-5

DETERMINE THE FUNCTIONAL RELATIONSHIP FOR THE FOLLOWING COST DATA TO MANUFACTURE PRODUCT X.

WHAT IS THE MEANING OF THE COST REPRESENTED BY THE Y-AXIS INTERCEPT ?

QUANTITY	5	10	15	20	25	30
COST	59	74	82	86	94	114

****

THE PROBLEM IS ONE OF FINDING THE LINEAR REGRESSION OF COST VERSUS QUANTITY.

$\Sigma X_i^2 = 105 \quad \Sigma X_i^2 = 2275 \quad \Sigma Y_i = 509 \quad \Sigma X_i Y_i = 9755$

$$\Sigma Y_i = aN + b\,\Sigma X_i \longrightarrow 509 = 6A + 105B$$
$$\Sigma X_i Y_i = a\,\Sigma X_i + b\,\Sigma X_i^2 \rightarrow 9755 = 105A + 2275B$$

$A = 50.93 \quad B = 1.937 \quad r = 0.974$

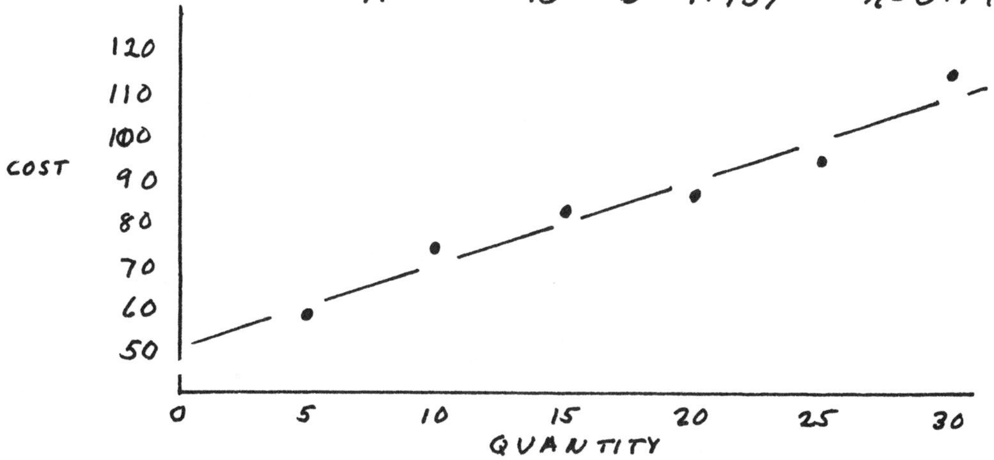

THE Y-AXIS INTERCEPT IS AN ESTIMATE OF THE FIXED COST.

10-6

Consider n pairs of data $(x_1, y_1), \ldots, (x_n, y_n)$ in which x denotes an independent variable and y denotes the dependent variable. Show that the estimated linear regression of y on x is given by the line

$$y - \bar{y} = m\,(x - \bar{x})$$

where

$$\bar{x} = \frac{1}{n}\sum_{i=1}^{n} x_i \qquad \bar{y} = \frac{1}{n}\sum_{i=1}^{n} y_i$$

and

$$m = \left[\frac{1}{n}\sum_{i=1}^{n}(x_i\,y_i) - \bar{x}\,\bar{y}\right] \left[\frac{1}{n}\sum_{i=1}^{n} x_i^2 - \bar{x}^2\right]$$

One way to estimate the capacitance C of an electric capacitor is to charge it to voltage v_o using a constant voltage source (switch in position 1) and then discharge it through a known resistance R (switch in position 2) while measuring the voltage decay. A schematic of this arrangement is given by the circuit shown.

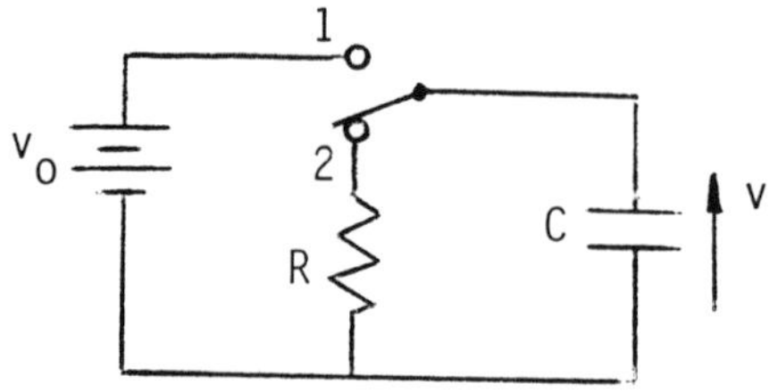

The voltage across the capacitor is known to obey the exponential relationship

$$v(t) = v_o\, e^{-t/(RC)}$$

during discharge. Three separate tests were conducted to measure the voltage decay. The same time increment (0.1 sec) and time interval (0.5 sec) were used in all three tests. The data is given below:

Time t (sec)		0.1	0.2	0.3	0.4	0.5
Voltage V (volts)	Test 1	7.3	2.8	1.0	0.4	0.1
	Test 2	7.4	2.7	1.1	0.3	0.2
	Test 3	7.3	2.6	1.0	0.4	0.1

If the resistance is precisely known to be 1000 Ohms, estimate the capacitance C in microFarads and the source voltage v_o in volts.

****

suppose the regression line is given by

$$y = mx + a \qquad \text{——— (i)}$$

The sum of squared measurement error is

$$e = \sum_{i=1}^{n} (y_i - mx_i - a)^2$$

NOTE: $mx_i + a$ is the y co-ordinate at x_i on the regression line and y_i is the actual measurement at x_i.

We must minimize e with respect to the parameters m and a.

Thus; $\frac{\partial e}{\partial m} = 0$ and $\frac{\partial e}{\partial c} = 0$ which give

$$\sum_{i=1}^{n} x_i (y_i - mx_i - a) = 0 \quad \text{and} \quad \sum_{i=1}^{n} (y_i - mx_i - a) = 0$$

Dividing these two equations by n we get

$$\frac{1}{n}\sum x_i y_i - \frac{m}{n}\sum x_i^2 - a\bar{x} = 0 \qquad \text{—— (ii)}$$

$$\bar{y} - m\bar{x} - a = 0 \qquad \text{—— (iii)}$$

Eliminate a by substituting (iii) in (ii). We get

$$m = \frac{\frac{1}{n}\sum_{i=1}^{n}(x_i y_i) - \bar{x}\bar{y}}{\frac{1}{n}\sum_{i=1}^{n}(x_i^2) - \bar{x}^2}$$

Subtract (iii) from (i);

$$y - \bar{y} = m(x - \bar{x})$$

For the numerical part of the problem;

$$V = V_0 \, e^{-t/(RC)}$$

Taking the natural log of the equation;

$$\ln V = -\frac{t}{RC} + \ln V_0$$

With $y = \ln V$ and $x = t$ this represents a straight line with slope $m = -\frac{1}{RC}$ and the y-intercept $\bar{y} - m\bar{x} = \ln V_0$.

For the given data

$\bar{x} = 0.3$, $\bar{y} = -0.01335$, $\frac{1}{n}\sum x_i y_i = -0.2067$

and $\frac{1}{n}\sum x_i^2 = 0.11$

NOTE: Take the log of the voltage data to get the corresponding y_i data.

Thus $m = -10.13 = -\frac{1}{RC}$

With $R = 1000$ we have $C = \frac{1}{10.13 \times 1000}$ Farad

or $C = 98.72\ \mu F$ $(1\ \mu F = 1 \times 10^{-6} F)$

Now $\ln V_0 = \bar{y} - m\bar{x} = -0.01335 + 10.13 \times 0.3$

Hence $V_0 = 20.61$ Volts

10-7

Analysis of weather precipitation records during the past winter reveals that the amount of precipitation, inches, that will fall tomorrow is a function of the center of a low pressure trough in miles from campus.

Precipitation, inches	0.2	0.4	0.6	0.8	1.0	1.4
Distance, 100 miles	37	22	16	13	11	8

Develop a precipitation prediction equation for the above data assuming $y = a + bx$ where x is the distance and y is precipitation.

****

$n = 6$

$\Sigma X = 107$ $\Sigma Y = 4.4$

$\bar{X} = 17.83$ $\bar{Y} = 0.733$

$(\Sigma x)^2 = 11{,}449 \qquad (\Sigma x)(\Sigma y) = 470.8$

$(\Sigma x)^2/n = 1908 \qquad (\Sigma x)(\Sigma y)/n = 78.467$

$\Sigma x^2 = 2463 \qquad \Sigma xy = 58.40$

$SS_x = 555 \qquad SP = -20.067$

where $SS_x = \Sigma xy - (\Sigma x)(\Sigma y)/n \qquad SP = \Sigma x^2 - (\Sigma x)^2/n$

$$b = \frac{SP}{SS_x} = \frac{-20.067}{555} = -0.036$$

$$a = \bar{y} - b\bar{x} = 0.733 - (-0.036)(17.83) = 1.375$$

$$y = 1.375 - 0.036X$$

10-8

Given the following data:

x	1	2	3	4	5	6
y	6	4	3	5	4	2

a) Plot on a scattergram. b) Find the regression line. c) Determine r^2 for the data.

****

(a)

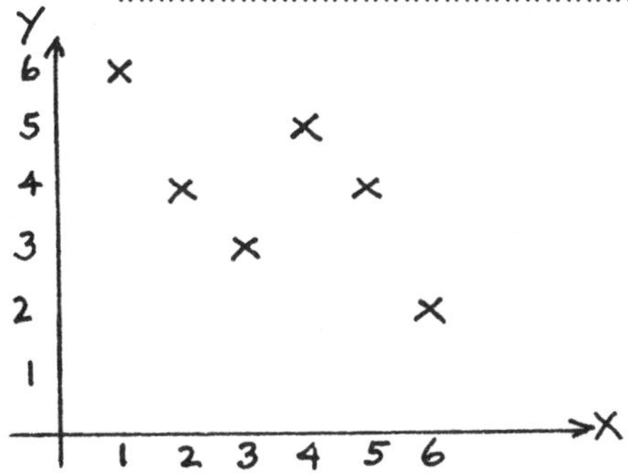

COMPUTE THE FOLLOWING:

$\Sigma x = 21 \qquad \Sigma y = 24 \qquad \Sigma XY = 75$

$\Sigma X^2 = 91 \qquad \Sigma y^2 = 106$

$$SS_{xx} = \Sigma x^2 - \frac{(\Sigma x)^2}{n} = 17.5$$

$$SS_{yy} = \Sigma y^2 - \frac{(\Sigma y)^2}{n} = 10$$

$$SS_{xy} = \Sigma xy - \frac{\Sigma x \Sigma y}{n} = -9$$

(b) then the regression line is $\hat{y} = a + bx$ where $b = \frac{SS_{xy}}{SS_{xx}} = \frac{-9}{17.5} = -.514$

$$a = \frac{\Sigma y}{n} - b\frac{\Sigma x}{n} = 5.799$$

AND $\hat{Y} = 5.799 - .514X$

(c) $$r^2 = \frac{(SS_{xy})^2}{SS_{xx} \cdot SS_{yy}} = \frac{(-9)^2}{(17.5)(10)} = 0.463$$

CONFIDENCE LIMITS

10-9

Analysis of weather precipitation records during the past winter reveals that the amount of precipitation, inches, that will fall tomorrow is a function of the center of a low pressure trough in miles from campus.

Precipitation, inches	0.2	0.4	0.6	0.8	1.0	1.4
Distance, 100 miles	37	22	16	13	11	8

Determine the 90 percent confidence limits for the regression coefficient, b, for the precipitation prediction equation for the above data if $y = a + bx$ where x is the distance and y is precipitation.

****

$n = 6$

$\Sigma x = 107$ $\Sigma y = 4.4$

$\bar{x} = 17.83$ $\bar{y} = 0.733$

$(\Sigma x)^2 = 11,449$ $(\Sigma x)(\Sigma y) = 470.8$ $(\Sigma y)^2 = 19.36$

$(\Sigma x)^2/n = 1908$ $(\Sigma x)(\Sigma y)/n = 78.467$ $(\Sigma y)^2/n = 3.23$

$\Sigma x^2 = 2463$ $\Sigma xy = 58.40$ $\Sigma y^2 = 4.16$

$SS_x = 555$ $SP = -20.067$ $SS_y = 0.93$

$$b = \frac{SP}{SS_x} = \frac{-20.067}{555} = -0.036$$

$$Reg\ SS = \frac{SP^2}{SS_x} = \frac{(-20.067)^2}{555} = 0.73$$

$$Res\ SS = SS_y - Reg\ SS = 0.93 - 0.73 = 0.20$$

$$S^2_{y/x} = \frac{Res\ SS}{n-k-1} = \frac{0.20}{6-1-1} = 0.05$$

$$S_b^2 = \frac{S^2_{y/x}}{SS_x} = \frac{0.05}{555} = 9 \times 10^{-5} \qquad S_b = 0.00949$$

$$\text{Confidence Limits} = b \pm t S_b = -0.036 \pm (2.132)(0.00949)$$

$$= -0.036 \pm 0.020 = \underline{\underline{-0.056 \text{ to } -0.016}}$$

CORRELATION

10-10

The scores of two athletes for long jumps, in inches, are given as follows.

Athlete X; jump length in inches: 120, 135, 128, 120, 137

Athlete Y; jump length in inches: 119, 128, 135, 124, 139

Determine the sample correlation. Also, indicate the 95% confidence interval for the population correlation coefficient.

****

Let $\alpha = X - \bar{X}$, and $\beta = Y - \bar{Y}$, and tabulate the computations as follows.

X	Y	α	β	$\alpha\beta$	α^2	β^2
120	119	−8	−10	80	64	100
135	128	7	−1	−7	49	1
128	135	0	6	0	0	36
120	124	−8	−5	40	64	25
137	139	9	10	90	81	100

$$\bar{X} = \frac{1}{n_x} \sum X = \frac{640}{5} = 128$$

$$\bar{Y} = \frac{1}{n_Y} \sum Y = \frac{645}{5} = 129$$

$$\sum (X - \bar{X})(Y - \bar{Y}) = \sum \alpha\beta = 203$$

$$\sum (X - \bar{X})^2 = \sum \alpha^2 = 258$$

$$\sum (Y - \bar{Y})^2 = \sum \beta^2 = 262$$

The sample correlation is:

$$r = \frac{\sum (X - \bar{X})(Y - \bar{Y})}{\sqrt{\sum (X - \bar{X})^2 \sum (Y - \bar{Y})^2}} = \frac{203}{\sqrt{(258)(262)}}$$

or, $r = 0.78$

The population correlation coefficient is read directly from the graph between sample correlation and population correlation for a given bivariate normal population for various sample sizes at 95% confidence. From such a graph, the 95% confidence interval for the population correlation coefficient is given as

$$-0.23 < r_p < 0.97$$

10-11

The following data relates the hardness to the tensile strength of a new alloy:

Tensile Strength (y)	2	4	6	8	10
Rockwell Hardness (x)	4	10	8	16	12

a) Plot a scatter diagram of the data.

b) Compute the correlation coefficient for the data.

c) Find the least squares linear regression equation for estimating tensile strength from hardness.

****

a)

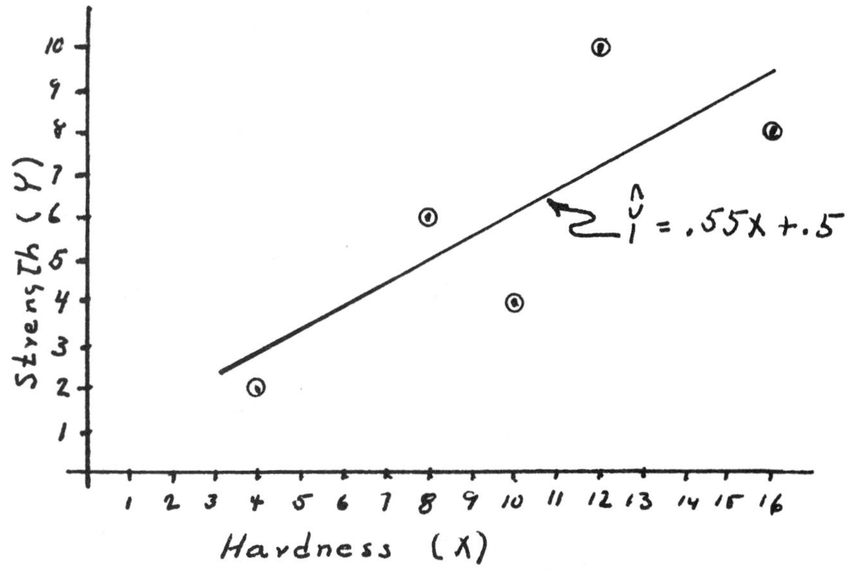

The following Computation table is developed

	X	Y	XY	X^2	Y^2
	4	2	8	16	4
	10	4	40	100	16
	8	6	48	64	36
	16	8	128	256	64
	12	10	120	144	100
Σ	50	30	344	580	220

b) $r = \dfrac{n\Sigma XY - \Sigma X \Sigma Y}{\sqrt{[n\Sigma X^2 - (\Sigma X)^2][n\Sigma Y^2 - (\Sigma Y)^2]}} = \dfrac{5(344) - (50)(30)}{\sqrt{[5(580) - (50)^2][5(220) - (30)^2]}}$

$= \dfrac{220}{\sqrt{80\,000}} = .778$

c) The equation is of the form $\hat{Y} = a + bx$

$a = \dfrac{\Sigma Y \Sigma X^2 - \Sigma X \Sigma XY}{n\Sigma X^2 - (\Sigma X)^2} = \dfrac{(30)(580) - (50)(344)}{5(580) - (50)^2} = .5$

$b = \dfrac{n\Sigma XY - \Sigma X \Sigma Y}{n\Sigma X^2 - (\Sigma X)^2} = \dfrac{5(344) - (50)(30)}{5(580) - (50)^2} = .55$

The prediction equation is $\hat{Y} = .55X + .5$

10-12

Analysis of weather precipitation records during the past winter reveals that the amount of precipitation, inches, that will fall tomorrow is a function of the center of a low pressure trough in miles from campus.

Precipitation, inches	0.2	0.4	0.6	0.8	1.0	1.4	y
Distance, 100 miles	37	22	16	13	11	8	x

What is the correlation between precipitation and distance?

****

$n = 6$

$\Sigma x = 107$ $\quad\quad \Sigma y = 4.4$

$\bar{x} = 17.83$ $\quad\quad \bar{y} = 0.733$

$(\Sigma x)^2 = 11,449$ $\quad (\Sigma x)(\Sigma y) = 470.8$ $\quad (\Sigma y)^2 = 19.36$

$(\Sigma x)^2/n = 1908$ $\quad (\Sigma x)(\Sigma y)/n = 78.467$ $\quad (\Sigma y)^2/n = 3.23$

$x^2 = 2463$ $\quad \Sigma xy = 58.40$ $\quad \Sigma y^2 = 4.16$

$SS_x = 555$ $\quad SP = -20.067$ $\quad SS_y = 0.93$

$r = \dfrac{SP}{\sqrt{SS_x\ SS_y}} = \dfrac{-20.067}{\sqrt{(555)(0.93)}} = \underline{\underline{-0.88}}$

10-13

Test the data below to see if y is a linear function of x at the 5% significance level:

x	y
0	3
1	6
2	14
3	35
4	82

****

If x and y are coded by subtracting the mean from each value (mean of the x's is 2, mean of y's is 28), the linear correlation coefficient r in terms of the coded x (say x') and coded y (y') is given by

$$r = \frac{\Sigma x'y'}{\sqrt{\Sigma(x')^2 \Sigma(y')^2}}$$

x	y	$x' = x-2$	$y' = y-28$	$(x')^2$	$(y')^2$	$x'y'$
0	3	-2	-25	4	625	50
1	6	-1	-22	1	484	22
2	14	0	-14	0	196	0
3	35	1	7	1	49	7
4	82	2	54	4	2,916	108
		0	0	10	4,270	187

Then $r = \frac{187}{\sqrt{(10)(4{,}270)}} = 0.905 > 0.878$ (The table r for 5% significance and 3 df*)

Conclusion: Correlation is significant between x and y according to a linear relationship, so it does appear that y is a linear function of x at 5% significance level.

* 3 df means 3 degrees of freedom obtained by subtracting the number of variables (2) from the number of points (5).

MULTIPLE REGRESSION

10-14

A marketing firm, in an attempt to determine an optimum mix of newspaper and radio advertising, conducted a test in which different amounts of money were spent on the two media to advertise a certain product in two similar market areas. The expenditures and the resulting sales are tabulated below.

Market Area	Radio $\$(10^{-2})$	Newspaper $\$(10^{-2})$	Gross Sales $\$(10^{-6})$
A	42	14	24
B	45	16	29
C	44	15	27
D	43	15	27
E	45	13	25
F	43	13	26
G	46	14	28
H	44	16	30
I	45	16	28
J	44	15	28

Perform a multi-variate least squares linear regression to develop an equation which will predict gross sales as a function of the amount spent on radio advertising and newspaper advertising.

****

The regression equation will be of the form

$$\hat{y} = a + b_1 X_1 + b_2 X_2$$

Where $\hat{y}$ = The estimated gross sales

X_1 = The amount spent on radio advertising

X_2 = the amount spent on newspaper advertising

a = the y intercept

b_1 & b_2 are the slopes associated with X_1 & X_2

The normal equation to be solved for a, b_1 & b_2 are

$$\Sigma Y = na + b_1 \Sigma X_1 + b_2 \Sigma X_2$$
$$\Sigma X_1 Y = a \Sigma X_1 + b_1 \Sigma X_1^2 + b_2 \Sigma X_1 X_2$$
$$\Sigma X_2 Y = a \Sigma X_2 + b_1 \Sigma X_1 X_2 + b_2 \Sigma X_2^2$$

By either computer or manual computation the following values are computed:

$\Sigma Y = 272$, $\Sigma X_1 = 441$ $\Sigma X_2 = 147$ $\Sigma X_1 Y = 12005$

$\Sigma X_2 Y = 4013$ $\Sigma X_1 X_2 = 6485$ $\Sigma X_1^2 = 19416$ $\Sigma X_2^2 = 2173$

Substituting into the normal equations gives

$$272 = 10a + 441b_1 + 147b_2$$

$$12005 = 441a + 19461b_1 + 6485b_2$$

$$4013 = 147a + 6485b_1 + 2173b_2$$

Solving simultaneously yields the following values

$$a = 13.828 \quad b_1 = .564 \quad b_2 = 1.099$$

and the required prediction equation is

$$\hat{Y} = 13.828 + .564X_1 + 1.099X_2$$

10-15

Determine if there is significant correlation between z and the variables x and y according to the relationship z = A + Bx + Cy (where A, B, and C are regression coefficients) for the following data at the 5% significance level. (z values are shown within the table.)

y \ x	10	20	30
3	61	102	144
4	57	110	155
5	60	101	164

****

Simplify calculations by coding the x, y, and z values by subtracting the mean (average) from each. Thus

$x' = x - 20$, $y' = y - 4$, $z' = z - 106$ (Prime denotes coded value)

Then $\Sigma x' = \Sigma y' = \Sigma z' = 0$, $\Sigma(x')^2 = 600^*$, $\Sigma(y')^2 = 6$, $\Sigma(z')^2 = 13{,}808$, $\Sigma x'y' = 0$, $\Sigma x'z' = 2{,}850$, $\Sigma y'z' = 18$

$$B = \frac{\Sigma x'z'}{\Sigma(x')^2} = \frac{2{,}850}{600} = \underline{4.75} \qquad C = \frac{\Sigma y'z'}{\Sigma(y')^2} = \frac{18}{6} = \underline{3.00}$$

Correlation coefficient r for this multiple linear regression =

$$\sqrt{\frac{(4.75)(2{,}850) + (3)(18)}{13{,}808}} = 0.992 > 0.795$$

(the table r value for degrees of freedom = 9 values of z minus 3 variables = 6 and 5% significance level)

Conclusion: Significant correlation. Thus, $z = A + Bx + Cy$ is a good predictor equation.

* This value of $\Sigma(x')^2 = 600$ may not be clear. Remember, this is obtained by adding the squares of the coded x values over 3 rows of the table; in other words,

$$\Sigma(x')^2 = 3\cdot\left[(-10)^2 + (0)^2 + (10)^2\right] = 600 \quad (\text{Similar for } \Sigma(y')^2.)$$

11
ANALYSIS OF VARIANCE

ONE - WAY

11-1

A completely randomized one-way analysis of variance (ANOVA) is to be performed with three treatment levels, given the response variable observations shown below. Perform the ANOVA at the 5% level of significance.

Treatment level 1: −5, −3, +2, 0, +1
2: +1, +1, +4, −1, +2, 0, +1
3: +1, −2, −1, −3, +2

****

$H_0: \mu_1 = \mu_2 = \mu_3$ $\qquad$ C = correction term = 0

H_1: H_0 not true

$$SST = \sum_{i=1}^{3} \sum_{j=1}^{n_i} y_{ij}^2 - C = 82 - 0 = 82.00$$

$$SS\,(Treat) = \left[\frac{(-5)^2}{5} + \frac{(+8)^2}{7} + \frac{(-3)^2}{5}\right] - 0 = 15.94$$

$$SSE = 82.00 - 15.94 = 66.06$$

Source	dF	SS	MS	F_{cal}		$F_{.05,2,14}$
Treat	2	15.94	7.97	1.69	<	3.74
Error	14	66.06	4.72			
Total	16	82.00				

Cannot reject H_0. Conclude that treatment has no significant effect on response variable.

11-2

In order to determine whether there is a difference in the strengths of welds made by four spot-welding machines at the K&S Machine Shop, the following readings were recorded. Each reading is the breaking force for the weld.

MACHINE	BREAKING FORCE					
I	27	31	17	13	9	20
II	37	29	28	19	20	24
III	23	19	22	18	15	24
IV	22	12	11	17	16	21

At a significance level of 0.01, should we reject the hypothesis that the mean pull force required to break the welds is the same for the four welders?

****

μ_i = MEAN BREAKING FORCE FOR MACHINE i ; $\alpha = 0.01$

$H_0: \mu_I = \mu_{II} = \mu_{III} = \mu_{IV}$; H_A : MEANS ARE NOT THE SAME

$df_{FACTOR} = 4-1 = 3$; $df_{ERROR} = 4(6-1) = 20$; $df_{TOTAL} = 24-1 = 23$

$SS_{TOTAL} = \Sigma x_i^2 - \dfrac{(\Sigma x_i)^2}{24}$; $SS_{FACTOR} = \dfrac{\Sigma R_i^2}{6} - \dfrac{(\Sigma x_i)^2}{24}$

(R_i = ROW SUM)

$SS_{ERROR} = \Sigma x_i^2 - \dfrac{\Sigma R_i^2}{6}$

$MS = SS/df$

SOURCE	df	SS	MS
FACTOR	3	295.17	98.39
ERROR	20	730.67	36.53
TOTAL	23	1025.84	

THE F TEST STATISTIC IS

$F = MS_{FACTOR}/MS_{ERROR}$

$= 98.39/36.53 = 2.69$

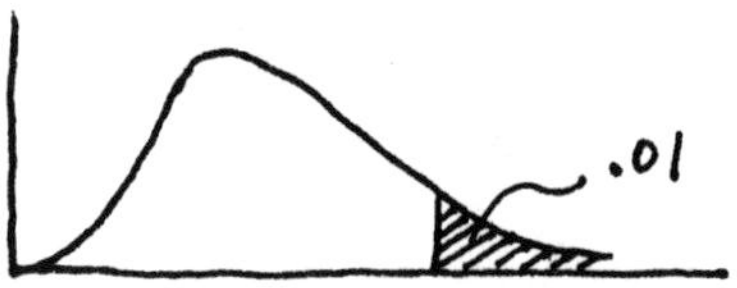

$F(3, 20, .01) = 4.94$

SINCE $2.69 < 4.94$, FAIL TO REJECT H_0 ; MEANS CAN BE TAKEN TO BE THE SAME.

11-3

Concrete cylinders were randomly selected from three batches of concrete, with each using a different mix design. Independent observations on compressive strength were made, with one observation coming from each cylinder. The data are as follows (with measurements in thousand psi):

Batch:	A	B	C
	5.6	5.9	6.4
	7.9	7.3	7.6
	7.7	6.8	
		7.6	
Totals:	21.2	27.6	14.0

Is there sufficient evidence to say that the mean strengths differ for the three batches? Use level of significance of 0.05.

****

$\bar{Y}_1 = 21.2/3 = 7.07, \quad R_1 = 2.3$

$\bar{Y}_2 = 27.6/4 = 6.90, \quad R_2 = 1.7$

$\bar{Y}_3 = 14.0/2 = 7.00, \quad R_3 = 1.2$

$\qquad\qquad\qquad 20.97 \qquad\qquad 5.2$

$\bar{\bar{Y}} = 20.97/3 = 6.99, \quad \bar{R} = 5.2/3 = 1.73$

$$TSS = \sum Y^2 - (\sum Y)^2/n = 443.88 - (62.8)^2/9$$
$$= 5.68$$

$$SST = \sum_{i=1}^{3} \frac{T_i^2}{n_i} - (\sum Y)^2/n = \sum_{i=1}^{3} \bar{Y}_i^2 n_i - (\sum Y)^2/n$$
$$= \frac{(21.2)^2}{3} + \frac{(27.6)^2}{4} + \frac{(14.0)^2}{2} - \frac{(62.8)^2}{9} = 0.05$$

$$SSE = TSS - SST = 5.68 - 0.05 = 5.63$$

$$F = \frac{MST}{MSE} = \frac{SST/(k-1)}{SSE/(n-k)} = \frac{0.05/2}{5.63/6} = 0.03$$

SINCE $F_6^2(0.05) = 5.14$, WE ACCEPT THE HYPOTHESIS THAT THE MEAN COMPRESSIVE STRENGTHS FOR THE THREE BATCHES ARE EQUAL.

11-4

A COMPANY IS THINKING ABOUT BUYING A NEW COMPUTER. FIVE VENDORS HAVE SUBMITTED BIDS THAT ARE CLOSE IN PRICE AND MEET THE SPECIFICATIONS.

TO DETERMINE WHICH ONE TO BUY, THE COMPANY ASKED EACH VENDOR TO RUN ONE OF THE COMPANY'S MAIN PROGRAMS WITH FIVE DIFFERENT DATA SETS AND TO REPORT THE COMPUTER RUN TIMES. THE RESULTS ARE AS FOLLOWS.

WHICH ONE SHOULD THEY BUY ?

	VENDOR				
	A	B	C	D	E
RUN					
1	57	42	30	45	56
2	60	46	43	49	52
3	45	53	35	58	45
4	62	56	37	53	48
5	57	58	25	41	52

****

THIS IS AN ANALYSIS OF VARIANCE PROBLEM

$T_i =$ 281 255 170 246 253 $T_. = 1205$

$$C = T_.^2/kN = 1205^2/25 = 58081$$

$$SST = \sum\sum Y_{ij}^2 - C = 60277 - 58081 = 2196$$

$$SS(Tr) = \sum T_i^2/N - C = 297411/5 - 58081 = 1401.2$$

$$SSE = SST - SS(TR) = 2196 - 1401.2 = 794.8$$

$$F = \frac{SS(Tr)/(k-1)}{SSE/k(N-1)} = \frac{(1401.2)/4}{(794.8)/20} = 8.81$$

$F_{4,4,.05} = 6.39 \rightarrow$ THERE IS A DIFFERENCE IN PERFORMANCE, SELECT C

$F_{4,4,.01} = 16.00 \rightarrow$ WE CANNOT DETECT A DIFFERENCE DO MORE TESTING

11-5

The following are percentages of hardwoods growing on four selected plots in a 40-acre tract of land owned by a lumber company:

Plot 1	Plot 2	Plot 3	Plot 4
25.6	25.2	20.8	31.6
24.3	28.6	26.7	29.8
27.9	24.7	22.2	34.3

Determine if there is significant difference among plots at the 1% level by using analysis of variance.

****

The plot totals (sums of each column in the table) are 77.8, 78.5, 69.7, and 95.7.
Sum of all 12 values is 321.7 (call this Σx).
Sum of squares of all 12 values is 8,788.81 (call this Σx^2).

Total sum of squares $SS_{tot} = 8{,}788.81 - \frac{(321.7)^2}{12} = 164.5692$ (with $8{,}788.81 \leftarrow \Sigma x^2$, $(321.7)^2 \leftarrow (\Sigma x)^2$)

Sum of squares between plots SS_{plot} =

$$\frac{(77.8)^2 + (78.5)^2 + (69.7)^2 + (95.7)^2}{3} - \frac{(321.7)^2}{12} = 119.6492$$

(3 = Number of values in each plot)

Sum of squares within plots (error) SS_{error} =

$$SS_{tot} - SS_{plot} = 164.5692 - 119.6492 = 44.9200$$

ANOVA table:

Source of Variance	SS		Degrees of Freedom		Mean Square (MS)
Between plots	119.6492	÷	3	=	39.883
Within plots (error)	44.9200	÷	8	=	5.615
Total	164.5692		11		–

Use the F test to compare mean square estimates and compare their ratio to the table F value for a 1% significance level (as specified) and degrees of freedom 3 (numerator) and 8 (denominator):

$$F = \frac{39.883}{5.615} = 7.103 < 7.59 \text{ (table value)}$$

Conclusion: No significant difference among plot percentages at 1% level.

11-6

A test is repeated m times independently. In each test n measurements are made. If x_{ij} denotes the jth measurement in the ith test, the following data values are obtained:

		Measurement				
		1	2	3	. .	n
	1	x_{11}	x_{12}	x_{13}		x_{1n}
Test	2	x_{21}	x_{22}	x_{23}		x_{2n}
	.					
	.					
	m	x_{m1}	x_{m2}	x_{m3}		x_{mn}

Determine an estimator for the variance of a measurement by considering

(i) variation within each test
(ii) variation between different tests

What is the degree of freedom in each of these estimators?

Production accuracies of three practically identical milling machines are compared by measuring the machining accuracy in a random sample of 5 gear wheels produced by each milling machine. The test data are given below:

		Measurement				
		1	2	3	4	5
	1	0.10	0.11	0.13	0.10	0.09
Machine	2	0.15	0.12	0.10	0.14	0.13
	3	0.09	0.11	0.08	0.10	0.08

Using one-way analysis of variance and an F test, check the hypothesis that the production accuracies of the three milling machines are identical at the 1% level of significance.

****

(i) The mean value from each test

$$\bar{x}_i = \frac{1}{n} \sum_{j=1}^{n} x_{ij}$$

The sample variance for each test

$$s_i^2 = \frac{\sum_{j=1}^{n} (x_{ij} - \bar{x}_i)^2}{(n-1)}$$

The average sample variance from all m tests

$$s_a^2 = \frac{\sum_{i=1}^{m} s_i^2}{m} = \frac{\sum_{i=1}^{m}\sum_{j=1}^{n}(x_{ij}-\bar{x}_i)^2}{m(n-1)}$$

s_a^2 is an estimator for the variance σ^2 of a measurement, considering variation within each test. This estimator has $m(n-1)$ degrees of freedom.

(ii) The mean value of all $m \times n$ measurements

$$\bar{x} = \frac{1}{mn}\sum_{i=1}^{m}\sum_{j=1}^{n} x_{ij}$$

An estimator for variance between "average" measurements from different tests is

$$\bar{s}^2 = \frac{\sum_{i=1}^{m}(\bar{x}_i - \bar{x})^2}{(m-1)}$$

Since n data values are averaged for each test, an estimator for the measurement variance σ^2 is

$$s_b^2 = n\bar{s}^2 = \frac{n}{(m-1)}\sum_{i=1}^{m}(\bar{x}_i - \bar{x})^2$$

This estimator has $(m-1)$ degrees of freedom.

For the given data, $m = 3$ and $n = 5$. Also

$\bar{x} = 0.1087$, $\bar{x}_1 = 0.106$, $\bar{x}_2 = 0.128$, $\bar{x}_3 = 0.092$

$\bar{s}_1^2 = 2.3 \times 10^{-4}$, $\bar{s}_2^2 = 3.7 \times 10^{-4}$, and $\bar{s}_3^2 = 1.7 \times 10^{-4}$

Hence

$\bar{s}_a^2 = 2.567 \times 10^{-4}$ with 12 d.o.f.

Also, $\bar{s}^2 = 3.2933 \times 10^{-4}$. Hence

$$s_b^2 = 5 \times 3.2933 \times 10^{-4} = 14.467 \times 10^{-4} \text{ with 2 d.o.f.}$$

$$\text{The F-ratio} = \frac{s_b^2}{s_a^2} = \frac{16.467 \times 10^{-4}}{2.567 \times 10^{-4}} = 6.42$$

At the 1% level of significance

$$F_{0.01,2,12} = 6.93 \quad \text{(From F tables)}$$

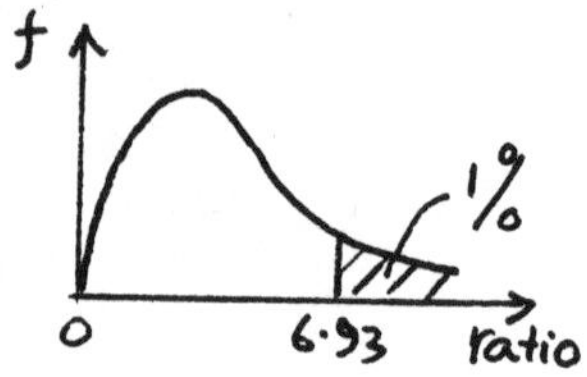

It follows that we can allow a F-ratio as large as 6.93 while accepting the hypothesis that the three machines are equally accurate. Now since the experimental value of the F-ratio 6.42 is less than the allowed upper limit 6.93 we accept the hypothesis at the 1% level of significance.

TWO-WAY

11-7

The following data was collected for a random block experiment:

BLOCK	TREATMENT 1	2	3
I	2	5	-1
II	-1	3	-1
III	-3	2	-3
IV	0	4	-1

Analyze the data and test the hypotheses: treatment means are equal
block means are equal

****

ROW TOTALS ARE 6, 1, −4, 3 $T_{i\cdot}$ $\Sigma T_{i\cdot}^2 = 62$

COL TOTALS ARE −2, 14, −6 $T_{\cdot j}$ $\Sigma T_{\cdot j}^2 = 236$

GRAND TOTAL IS 6 $T_{\cdot\cdot}$ $T_{\cdot\cdot}^2 = 36$

$\Sigma y_{ij}^2 = 80$

CORRECTION FACTOR (CF) $= T_{\cdot\cdot}^2/n = 36/12 = 3$

then $SS_{TOTAL} = 80 - 3 = 77$

$SS_{TREAT} = \frac{236}{4} - 3 = 56$

$SS_{BLOCK} = \frac{62}{3} - 3 = 17.67$

$SS_{ERROR} = 77 - 56 - 17.67 = 3.33$

CONSTRUCT THE ANOVA TABLE:

SOURCE	SS	df	MS	F	$F_{.01}$
TREATMENTS	56	2	28	50	10.92
BLOCKS	17.67	3	5.89	10.52	9.78
ERROR	3.33	6	0.56		
TOTAL	77	11			

therefore both Treatments and blocks are significant at the .01 level: both hypotheses are rejected.

MULTIPLE

11-8

The forestry division of a large forest products manufacturer wanted to compare growth rates of three hardwoods - post oak (O), sweet gum (G), and hickory (H) - under three different soil conditions - clay (C), sand (S), and loam (L). Three growth plots were staked out for each soil type. On each of these three plots several trees, all of the same type, were planted. One year later the diameters of all trees on a given plot were measured and an average diameter calculated. After another five years, the same trees were again measured and an average diameter calculated for each plot. Then the percent increase in average diameter was calculated for each plot for the five years of growth. These results are below:

		Soil Type C	Soil Type S	Soil Type L
Tree Type	O	205 (Tableland plot)	210 (Swamp plot)	365 (Ridge plot)
	G	320 (Swamp plot)	350 (Ridge plot)	440 (Tableland plot)
	H	180 (Ridge plot)	160 (Tableland plot)	250 (Swamp plot)

An analysis of variance on the influence of tree type and soil type on the percent increases in growth shown within the table yielded the following:

Source	Sum of Squares	Degrees of Freedom	Mean Square
Trees	46,155.6	2	23,077.8
Soils	26,105.6	2	13,052.8
Error	2,711.1	4	677.8
Total	74,972.2	8	-

It was then realized that moisture might also be contributing to differences in growth, depending on plot location in swamp, ridge, or tableland. Test for significance of this moisture (location) effect only at the 5% significance level and state your conclusion.

****

The location sum of squares (with 2 degrees of freedom) is found by first totalling the percentages for each type of location:

Tableland = 205 + 160 + 440 = 805
Swamp = 320 + 210 + 250 = 780
Ridge = 180 + 350 + 365 = 895
total = 2,480

Then location sum of squares $= \dfrac{(805)^2+(780)^2+(895)^2}{3} - \dfrac{(2{,}480)^2}{9} = 2{,}438.89$

Error sum of squares is then = 2,711.1 (from table) − 2,438.89 = 272.22, reducing its degrees of freedom to 2.

F test: $F = \dfrac{\frac{2{,}438.89}{2}}{\frac{272.2}{2}} = 8.96 < 19.00$ (table F for degrees of freedom 2 and 2 and 5% significance)

Conclusion: Moisture effect does not significantly affect growth.

12 FACTORIAL EXPERIMENTS

12-1

The following table shows the number of square yards of carpet with dyeing defects among three dyeing lines and four shifts during a two-month period of production in a carpet mill. Test for a significant difference among lines, shifts, and interaction between lines and shifts at the 5% significance level.

	Shift			
Dyeing Line	A	B	C	D
I	302	94	1,147	256
	134	105	770	0
II	0	0	321	0
	126	91	362	0
III	40	53	229	0
	0	111	125	220

****

The sum of squares (SS) values for this 2-factor factorial design follow:

$$SS_{total} = \Sigma x^2 - \frac{(\Sigma x)^2}{24} = 2,494,504 - \frac{(4,486)^2}{24} = 1,655,995.83$$

$$SS_{line} = \frac{(2,808)^2 + (900)^2 + (778)^2}{8} - \frac{(4,486)^2}{24} = 324,010.33 \text{ (using row totals)}$$

$$SS_{shift} = \frac{(602)^2 + (454)^2 + (2,954)^2 + (476)^2}{6} - \frac{(4,486)^2}{24} = 748,360.50 \text{ (using column totals)}$$

$$SS_{interaction} = \frac{(436)^2 + (199)^2 + (1,917)^2 + \cdots + (220)^2}{2} - \frac{(4,486)^2}{24} - SS_{line} - SS_{shift} = 420,611.00 \text{ (using cell totals)}$$

$$SS_{error} = SS_{total} - SS_{line} - SS_{shift} - SS_{interaction} = 163,014.00$$

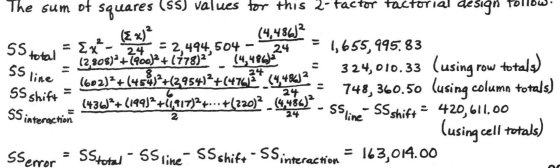

The F test is used to compare each mean square (MS) estimate to the error MS, then the result is compared to the table F value at the appropriate degrees of freedom (df) for the numerator and denominator of the F ratio (at 5% significance), as shown in the ANOVA table:

Source of Variance	SS	df	MS	F(calc.)		F(table)
Lines (L)	324,010.33	2	162,005.17	11.93	>	3.89
Shifts (S)	748,360.50	3	249,453.50	18.36	>	3.49
Interaction L×S	420,611.00	6	70,101.83	5.16	>	3.00
Error	163,014.00	12	13,584.50	–		–
Total	1,655,995.83	23	–	–		–

Conclusion: There is a significant difference among lines and among shifts as to square yards of dye defects produced, and there is a significant interaction effect between lines and shifts.

12-2

An instrument manufacturer uses a two terminal semiconductor device for temperature compensation of its pressure transducers. It is suspected that the resistance of the device is a function of impressed voltage, and hence, current. A 3**2 factorial design is used to test this suspicion. Three equally spaced levels of current and three equally spaced levels of resistance were used. Two replications were obtained by selecting devices with identical nominal resistances. All tests were run with the device skin temperature held at 25° C. The following voltages were observed:

Current (mA)	Nominal Resistance (ohms) 4000	5000	6000
0.5	2.729 2.549	3.207 3.310	3.328 3.547
1.0	4.702 4.752	5.586 5.824	7.050 7.095
1.5	6.836 6.869	8.805 8.864	10.751 9.925

Is there a non-zero voltage coefficient of resistance?

****

Form the cell totals $y_{ij\cdot}$, the row sums $y_{i\cdot\cdot}$, the column sums $y_{\cdot j\cdot}$, and the grand total $y_{\cdot\cdot\cdot}$

	4000	5000	6000	$y_{i\cdot\cdot}$
0.5	5.278	6.517	6.875	18.670
1.0	9.454	11.410	14.145	35.009
1.5	13.705	17.669	20.676	52.050
$y_{\cdot j\cdot}$	28.437	35.596	41.696	105.729

$$SS_T = \sum_{i=1}^{a}\sum_{j=1}^{b}\sum_{k=1}^{n} y_{ijk}^2 - \frac{y_{\cdot\cdot\cdot}^2}{abn} \qquad a = b = 3 \qquad n = 2$$

$$= (2.729)^2 + (2.549)^2 + \cdots + (9.925)^2 - \frac{(105.729)^2}{18}$$

$$= 111.764$$

$$SS_R = \sum_{i=1}^{a} \frac{y_{i..}^2}{bn} - \frac{y_{...}^2}{abn}$$

$$= \frac{(18.670)^2 + (35.009)^2 + (52.050)^2}{6} - \frac{(105.729)^2}{18}$$

$$= 92.866$$

$$SS_I = \sum_{j=1}^{b} \frac{y_{.j.}^2}{an} - \frac{y_{...}^2}{abn}$$

$$= \frac{(28.437)^2 + (35.596)^2 + (41.696)^2}{6} - \frac{(105.729)^2}{18}$$

$$= 14.681$$

$$SS_{RI} = \sum_{i=1}^{a} \sum_{j=1}^{b} \frac{y_{ij.}^2}{n} - \frac{y_{...}^2}{abn} - SS_R - SS_I$$

$$= \frac{(5.278)^2 + \cdots + (20.676)^2}{2} - \frac{(105.729)^2}{18} - 92.866$$

$$- 14.681 = 3.798$$

$$SS_E = SS_T - SS_R - SS_I - SS_{RI}$$

$$= 111.764 - 92.866 - 14.681 - 3.798 = 0.419$$

Source of Variation	Sum of Squares	Degrees of Freedom	Mean Square	F_0
Resistance	92.866	2	46.433	997
Current	14.681	2	7.341	158
Interaction	3.788	4	0.950	20.4
Error	0.419	9	0.047	
Total	111.764			

$F_{.001,4,9} = 12.56$ therefore $F_0 = 20.4$ is significant at the 0.1% level and we conclude that the interaction between resistance and current is responsible for part of the variation in observed voltage. The voltage coefficient of resistance is non-zero.

13 NONPARAMETRIC STATISTICS

SIGN TEST

13-1

Listed below are the scores on a standardized engineering examination for eight students selected at random from all students who have completed the exam. Test the hypothesis (at the 5% level of significance) that the mean score for all students taking the exam is 65, against the alternative that the mean is less than 65. Several goodness of fit tests have failed to justify any assumptions about the probability density function of the population.

67,76,60,88,32,68,72,70

****

$H_0: \mu = 65$ or $H_0: p = 0.50$ where $p = P(+)$

$H_1: \mu < 65$ $H_1: p < 0.50$ $+$ = value > 65

$-$ = value < 65

Let x = no. of "+" values = 6

Note that obviously H_0 cannot be rejected since 6 of 8 values exceed 65. However, the test could be continued as follows using the sign test:

The binomial distribution is appropriate with parameters $n = 8$, $p = 0.50$.

Reject H_0 if $P(X \le 6) < \alpha = 0.05$

$P(X \le 6) = 0.9648 \not< 0.05$. Cannot reject H_0. There is insufficient evidence that $\mu < 65$. In fact, $\mu > 65$ is likely.

(Binomial table)

13-2

From a large student class a random sample of 5 students with grades, 62, 75, 80, 87, and 92 was taken for investigative purposes. If the median grade for the whole class is reported to be 76, indicate if the class median grade is actually higher than 76.

****

The Null Hypothesis H_0: Median = 76

The Alternative Hypothesis H_1: Median > 76

Since, half the class grades lie above 76, therefore, if a grade is drawn randomly, the probability that it lies above 76 is:

$$H_0 : \pi = \frac{1}{2}$$

Then, if H_0 is true, the sample of $n = 5$ is just like tossing a coin 5 times. The total number of Successes S_0 (grades above 76) will have the binomial distribution. Then for $n = 5$, $S_0 = 3$ (grades above 76), and $\pi = \frac{1}{2}$, the probability of successes from the binomial probability tables is given as 0.5000. This probability is so high at the 5% level that H_0 cannot be rejected. Therefore, it is concluded that H_0 is true and the class median grade is not higher than 76.

13-3

To compare the accuracies of two types of displacement transducers (say a potentiometer and a variable differential transformer) the same rotation was measured (in degrees) using the two devices. Ten such measurement pairs were taken;

Pot	10.3	5.6	20.1	15.2	2.0	7.6	12.1	18.9	22.1	25.2
VDT	9.8	5.5	20.0	16.0	1.9	7.4	12.2	18.7	22.0	25.0

Are the two transducers equally accurate at the 5% level of significance? Use a sign test.

****

From the given data;

Pot-VDT	0.5	0.6	0.1	-0.8	0.1	0.2	-0.1	0.2	0.1	0.2

Note that there are 2 negative differences and 8 positive differences. But if we used VDT-Pot we would get 8 negative differences and 2 positive differences. Both these cases should be considered in the sign test.

If the two devices are equally accurate

P[negative difference] = P[positive difference] = 0.5

The number of differences with the same sign in a set of measurements obeys the binomial distribution. For the given data

P[2 or less negative signs or 8 or more negative signs]

$$= {}^{10}C_0(0.5)^{10} + {}^{10}C_1(0.5)^{10} + {}^{10}C_2(0.5)^{10} + {}^{10}C_8(0.5)^{10} + {}^{10}C_9(0.5)^{10} + {}^{10}C_{10}(0.5)^{10}$$

$$= 2 \times (0.5)^{10}\left({}^{10}C_0 + {}^{10}C_1 + {}^{10}C_2\right) = 0.11$$

{ Assuming that the two devices are equally accurate

This probability (11%) is greater than the allowed lower limit (5%). It is concluded the two devices are equally accurate at the 5% level of significance.

WILCOXON TWO-SAMPLE RANK TEST

13-4

Two techniques are being tested for measuring the diameters of glass filaments which are used in optical communications. One technique is mechanical which is relatively slow; the other is optical sensoring which is quite fast. The diameter is supposed to be 10.00 mm. Thirty identifiable filaments were mixed in with 970 good filaments and measured by both techniques. The diameters are given below. Since the size of the differences in diameters is important, use a 95% confidence level to determine if the two techniques are equally accurate.

	DIAMETER (mm)			DIAMETER (mm)	
FILAMENT	OPTICAL	MECHANICAL	FILAMENT	OPTICAL	MECHANICAL
1	9.92	9.96	16	9.96	9.91
2	9.82	9.80	17	9.80	9.88
3	10.04	9.99	18	10.01	10.10
4	9.95	9.95	19	10.10	10.00
5	10.01	10.00	20	9.89	9.89
6	9.90	9.96	21	9.91	10.04
7	10.08	10.07	22	9.95	9.96
8	9.89	9.96	23	10.00	9.89
9	10.00	10.03	24	9.80	9.80
10	10.04	9.99	25	9.99	10.08
11	9.89	9.91	26	10.00	10.01
12	9.97	10.03	27	9.85	9.85
13	9.86	9.99	28	9.92	9.98
14	10.03	9.91	29	10.02	9.96
15	9.84	9.87	30	10.04	10.00

****

$H_0: \mu_o = \mu_m$; $H_A: \mu_o \neq \mu_m$; $\alpha = 0.05$

$d_i = x_i^o - x_i^m$

d_i	$\lvert d_i \rvert$	RANK		d_i	$\lvert d_i \rvert$	RANK		d_i	$\lvert d_i \rvert$	RANK	
	ORDER	PLUS	MINUS		ORDER	PLUS	MINUS		ORDER	PLUS	MINUS
−.04	.01		2.5	−.02	.05	12		−.13	.09		20.5
.02	.01		2.5	−.06	.05	12		−.01	.10	22	
.05	.01	2.5		−.13	.05	12		.11	.11	23	
.00	.01	2.5		.12	.06		15.5	.00	.12	24	
.01	.02		5.5	−.03	.06		15.5	−.09	.13		25.5
−.06	.02	5.5		.05	.06		15.5	−.01	.13		25.5
.01	.03		7.5	−.08	.06	15.5		.00	—		
−.07	.03		7.5	−.09	.07		18	−.06	—		
−.03	.04		9.5	.10	.08		19	.06	—		
.05	.04	9.5		.00	.09		20.5	.04	—		
										140.5,	210.5

THE TEST STATISTIC IS $T = \min(140.5, 210.5) = 140.5$.
SINCE $n = 26$, USE

$$z = \frac{T - \mu_T}{\sigma_T}, \quad \mu_T = \frac{n(n+1)}{4}, \quad \sigma_T = \sqrt{\frac{n(n+1)(2n+1)}{24}}$$

$$\mu_T = \frac{26 \times 27}{4} = 175.5 \; ; \quad \sigma_T = \sqrt{\frac{26 \times 27 \times 53}{24}} = 39.37$$

$$z = \frac{140.5 - 175.5}{39.37} = -0.89$$

FOR $\alpha = .05$ WITH A TWO-SIDED TEST, THE ACCEPTANCE REGION IS $|z| \leq 1.96$.

SINCE $-0.89 > -1.96$, FAIL TO REJECT H_0. THE TWO TECHNIQUES ARE EQUALLY ACCURATE.

RUNS TEST

13-5

The flow rate of water in a pipe comprising part of a cooling system must be closely controlled. The prescribed rate is 1.5 gal/sec. This value was used as a reference to determine the following sequence of readings where a = above and b = below:

abbaaababbaabbbababb.

Do these readings represent a random sample at the 5% significance level?

****

H_0: SAMPLE IS RANDOM ABOUT 1.5 gal/sec
H_A: SAMPLE IS NOT RANDOM

$\alpha = 0.05$, n_1 = # BELOW = 11, n_2 = # ABOVE = 9

r = # RUNS = 12 THIS IS THE TEST STATISTIC.

FROM THE r-DISTRIBUTION, $6 \leq r \leq 16$ IS THE ACCEPTANCE REGION. DO NOT REJECT H_0.

13-6

A sample of 35 observations of a random variable has been collected for a nonparametric sign test. However, before the sign test is performed, we want to test whether the sample can be considered to be a random sample. The sample data is given below in the order in which it was collected (reading left to right), with observations equal to the median of 8 discarded. Test for randomness using a 5% level of significance.

9,10,6,9,7,12,15,7,6,9,11,10,10,7,6,7,11,6,10,6,7,11,10,5,9,7,10,6,9,12

****

Nonparametric Runs test is appropriate

H_0: Sample is random $\qquad z_{.025} = 1.96$
H_1: H_0 not true

Let "+" = value > 8 $\qquad n_1$ = no. of "+" values = 17
"−" = value < 8 $\qquad n_2$ = no. of "−" values = 13

u = no. of runs = 19

$$\mu_u = \frac{(2)(17)(13)}{30} + 1 = 15.733$$

$$\sigma_u = \sqrt{\frac{(2)(17)(13)[(2)(17)(13) - 17 - 13]}{(30)^2(29)}} = 2.641$$

$$z_{cal} = \frac{19 - 15.733}{2.641} = 1.237 < z_{.025} = 1.96,$$

cannot reject H_0; conclude sample random

OTHER TOPICS

13-7

A random sample of 30 service times was taken at a local fast food restaurant, and are given below. Test these data to see if there is evidence that service times may not be considered to have an exponential distribution. Use a significance level of 5%.

Observed Service Times (seconds)

17	45	54	27	19	29
10	58	15	18	7	12
44	13	57	57	43	53
35	9	39	12	29	55
56	45	19	57	22	34

****

DATA PT	VAL	TH	OBS	|D|	DATA PT	VAL	TH	OBS	|D|
1	7	.191	.033	.158	16	34	.643	.533	.110
2	9	.239	.067	.172	17	35	.654	.567	.087
3	10	.261	.100	.161	18	39	.693	.600	.093
4	12	.305	.167	.138	19	43	.728	.633	.095
5	12	.305	.167	.138	20	44	.736	.667	.069
6	13	.326	.200	.126	21	45	.744	.733	.011
7	15	.365	.233	.132	22	45	.744	.733	.011
8	17	.403	.267	.136	23	53	.799	.767	.032
9	18	.420	.300	.120	24	54	.805	.800	.005
10	19	.438	.367	.071	25	55	.811	.833	.022
11	19	.438	.367	.071	26	56	.817	.867	.050
12	22	.487	.400	.087	27	57	.822	.967	.145
13	27	.559	.433	.126	28	57	.822	.967	.145
14	29	.585	.500	.085	29	57	.822	.967	.145
15	29	.585	.500	.085	30	58	.828	1.00	.172

$\bar{x} = 33 \qquad TH = 1 - e^{-x/33}$

MAX $|D| = 0.172 \qquad CV = 0.24$

Cannot reject the hypothesis that the distribution is exponential, because the test statistic of 0.172 is less than the critical value of 0.24

14 QUALITY CONTROL

CONTROL CHARTS: MEAN, X-BAR

14-1

Thirty samples, each of size five, were drawn from a manufacturing process for quality control purposes. For a particular measurement of interest, the sum of all observations is 4050. The range of each sample was computed, yielding an average sample range of 5.20. Calculate upper and lower limits for a trial "three sigma" mean control chart and estimate the process standard deviation.

****

$$\bar{\bar{X}} = \frac{4050}{(30)(5)} = 27.00 \qquad \bar{R} = 5.20$$

$$UCL_{\bar{X}} = \overset{\bar{\bar{X}}}{27.00} + \overset{A_2}{(0.577)}\overset{\bar{R}}{(5.20)} = 30.00$$

$$LCL_{\bar{X}} = \overset{\bar{\bar{X}}}{27.00} - \overset{A_2}{(0.577)}\overset{\bar{R}}{(5.20)} = 24.00$$

$$\hat{\sigma}_x = \frac{\bar{R}}{d_2} = \frac{5.20}{2.326} = 2.236$$

14-2

The voltage needed to drive a computer chip must be closely controlled. Samples of size three were taken for a period of 12 days. The sample means and ranges were computed. Use the data shown to obtain control of the output. Find the tenative $\overline{X}$ control chart limits for the next period.

SAMPLE	$\overline{X}$	R	SAMPLE	$\overline{X}$	R
1	6.04	0.022	7	5.92	0.018
2	6.12	0.019	8	6.01	0.028
3	5.91	0.019	9	5.94	0.022
4	6.10	0.029	10	6.00	0.017
5	5.97	0.024	11	6.08	0.018
6	5.99	0.021	12	6.03	0.036

****

$k = 12$ = # OF RANDOM SAMPLES ; $n = 3$ = SIZE OF EACH SAMPLE

$\overline{\overline{X}}$ = AVERAGE OF THE MEANS = $72.11/12 = 6.009$

$\overline{R}$ = AVERAGE OF THE RANGES = $0.273/12 = 0.023$

$$CL_{\overline{X}} = \overline{\overline{X}} \pm A_2\overline{R} \quad , \quad A_2 = 1.023$$

$$= 6.009 \pm (1.023)(0.023) = 6.032 \text{ AND } 5.986$$

$$UCL_R = D_4\overline{R} = (2.574)(.023) = 0.059$$

$$LCL_R = D_3\overline{R} = 0(.023) = 0$$

ALL OF THE R VALUES ARE IN CONTROL.

ONLY SAMPLES 6, 8, 10, AND 12 HAVE $\overline{X}$ VALUES IN CONTROL.

H_0: PROCESS IN CONTROL, MUST BE REJECTED.

REVISE LIMITS USING SAMPLES 6, 8, 10, 12 : $k = 4$

$\overline{\overline{X}} = (5.99 + 6.01 + 6.00 + 6.03)/4 = 6.008$

$\overline{R} = (.021 + .028 + .017 + .036)/4 = 0.026$

$$CL_{\overline{X}} = 6.008 \pm (1.023)(.026) = 6.034 \text{ AND } 5.982$$

$$UCL_R = (2.574)(0.026) = 0.067 \; ; \; LCL_R = 0$$

NOW THE SELECTED SAMPLES ARE IN CONTROL (BOTH $\bar{X}$ AND R).

H_0 : PROCESS IN CONTROL ; DO NOT REJECT H_0 WITH THE REVISION.

FOR THE NEXT PERIOD : $CL_{\bar{X}} = 5.982$ AND 6.034

$CL_R = 0$ AND 0.067

THESE ARE NOT RELIABLE ESTIMATES BECAUSE MOST OF THE INITIAL SAMPLES WERE OUT OF CONTROL.

14-3

A set of N = 20 samples of size n = 6 is such that $\bar{\bar{x}}$ = 44.26 and $\bar{R}$ = 4.26. Assuming that the individuals are normally distributed and $\bar{x}$ & R charts show statistical controls, at what values should x_{min} and x_{max} be set if one can tolerate 3% of the items above x_{max} and 1% of the items below x_{min}.

****

σ' (ESTIMATED) = $\bar{R}/d_2 = 4.26/2.534 = 1.68$

$X_{MAX} = 44.26 + 1.881\,(1.68) = 47.42$

$X_{MIN} = 44.26 - 2.326\,(1.68) = 40.35$

14-4

Error in a satellite tracking system was monitored on line for a period of 1 hr, to determine whether recalibration of the control system is required. Four measurements were taken in a period of 5 minutes and 12 such data groups were acquired during the hour. Sample means and variances of the 12 groups of data were computed to be as follows:

Period i	1	2	3	4	5	6	7	8	9	10	11	12
Sample Mean $\bar{x}_i$	1.34	1.10	1.20	1.15	1.30	1.12	1.26	1.10	1.15	1.32	1.35	1.18
Sample Variance s_i^2	0.11	0.02	0.08	0.10	0.09	0.02	0.06	0.05	0.08	0.12	0.03	0.07

Draw a $\bar{X}$ chart for the error process, with action lines at $\bar{x} \pm 3\sigma$. Establish whether the control system needs recalibration.

****

The overall mean $\bar{\bar{x}} = \frac{1}{12}\sum_{i=1}^{12} \bar{x}_i = 1.214$

The average sample variance $\bar{s}^2 = \frac{1}{12}\sum_{i=1}^{12} s_i^2 = 0.069$

Since there are 4 measurements in each period, the variance of group mean $\bar{x}_i$ can be estimated as

$$\sigma^2 \simeq s^2 = \frac{\bar{s}^2}{4}$$

An estimator for the std. deviation σ of the sample means is

$$\sigma \simeq \frac{\bar{s}}{\sqrt{4}} = \frac{\sqrt{0.069}}{2} = 0.131$$

The upper action line is given by $x = \bar{x} + 3\sigma$

$$= 1.214 + 3 \times 0.131 = 1.607$$

The lower action line is $x = \bar{x} - 3\sigma = 0.821$

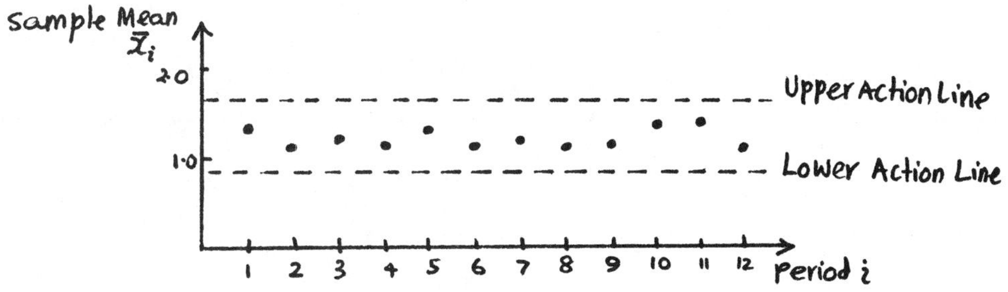

Since the sample means lie within the two action lines, the process is under control and no recalibration is necessary.

14-5

Twenty-eight samples of five each were drawn from a process. Construct an X-bar chart given:

$\Sigma\bar{X}=2216.0$ $\Sigma R=348.0$

****

$$\bar{\bar{X}} = \frac{2216}{28} = 79.14 \qquad \bar{R} = \frac{348}{28} = 12.43$$

FROM A TABLE OF CONTROL CHART CONSTANTS FIND A_2 (FOR A SAMPLE SIZE OF 5)

$$A_2 = .577$$

$$UCL = 79.14 + .577\,(12.43) = 86.31$$
$$LCL = 79.14 - .577\,(12.43) = 71.97$$

14-6

Each day over a 14-day period the machine speeds in feet per minute were recorded for 5 random observations during the day of carpet emerging from a drying oven in a carpet mill. The results are below. Determine if the sample means are in control according to an appropriate control chart.

Day	Sample Speeds				
1	103	127	118	127	127
2	127	112	109	100	109
3	100	127	100	109	106
4	103	103	100	106	103
5	91	100	97	100	88
6	79	106	100	100	106
7	91	97	97	100	94
8	100	94	91	91	94
9	106	100	97	91	97
10	100	94	97	94	94
11	91	85	115	100	115
12	82	91	100	100	76
13	91	94	97	97	106
14	97	97	97	94	97

****

The sample means for days 1-14 are as follows:

120.4, 111.4, 108.4, 103.0, 95.2, 98.2, 95.8, 94.0, 98.2, 95.8, 101.2, 89.8, 97.0, 96.4

The mean $\bar{\bar{X}}$ of the sample means (and central line for the control chart) is

$$\bar{\bar{X}} = \frac{\Sigma \text{ means}}{14} = \frac{1404.8}{14} = 100.3$$

Control limits are $\bar{\bar{X}} \pm A_2 \cdot \bar{R}$, with factor $A_2 = 0.58$ taken from a standard table for sample size 5, and $\bar{R}$, the average range of samples, calculated from the sample ranges as $\bar{R} = (24+27+27+6+\cdots+3) \div 14 = 16.7$
Thus, $UCL = 100.3 + (0.58)(16.7) = 110.0$, $LCL = 100.3 - (0.58)(16.7) = 90.6$.
The chart below shows 3 points (days) outside the control limits and a downward shift in the average speed after Day 4 (investigate causes):

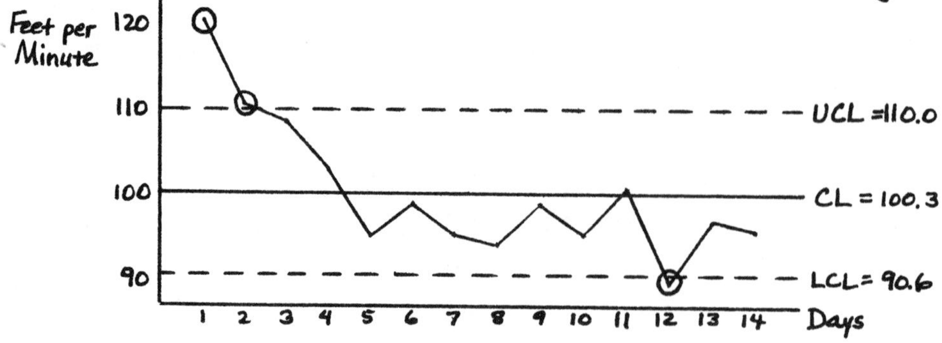

CONTROL CHARTS: RANGE, R

14-7

Bolts for oil pipe flanges are received in lots of 1000. The specified AQL = 6.5 percent.

(a) What is the single sampling plan that should be used? The general inspection level II is to be used. See Tables I-X of MIL-STD-105D, or equivalent, for single sampling plans for normal inspection.

(b) What is the probability of acceptance for this plan if the lots submitted are at AQL quality(6.5%)?
Hint: Use straight line interpolation for values in Table X of MIL-STD-105D, or equivalent.

****

(a) FROM TABLE I (MIL-STD-105D) LOT SIZE OF 1000 GIVES LETTER CODE J FOR GENERAL INSPECTION LEVEL II. FROM TABLE II (MIL-STD-105D) FOR LETTER CODE J THE SAMPLE SIZE 80 IS OBTAINED, AND $A_c = 10$ AND $R_e = 11$.

(b)

P_a	p
99.00	5.98
P_a	6.50
95.00	7.40

$$P_a = 95.0 + (99.0 - 95.0)\left(\frac{7.40 - 6.50}{7.40 - 5.98}\right)$$

$$= 97.54$$

14-8

Each day over a 14-day period the machine speeds in feet per minute were recorded for 5 random observations during the day of carpet emerging from a drying oven in a carpet mill. The results are below. Determine if the sample ranges are in control according to an appropriate control chart.

Day	Sample Speeds				
1	103	127	118	127	127
2	127	112	109	100	109
3	100	127	100	109	106
4	103	103	100	106	103
5	91	100	97	100	88
6	79	106	100	100	106
7	91	97	97	100	94
8	100	94	91	91	94
9	106	100	97	91	97
10	100	94	97	94	94
11	91	85	115	100	115
12	82	91	100	100	76
13	91	94	97	97	106
14	97	97	97	94	97

****

The sample ranges for Days 1-14 are as follows:

24, 27, 27, 6, 12, 27, 9, 9, 15, 6, 30, 24, 15, 3

The average range $\bar{R}$ (and central line for the control chart) is

$\bar{R} = \frac{\Sigma \text{ ranges}}{14} = \frac{234}{14} = 16.7$. Using standard tables, the upper and lower control limits are given as $D_4 \cdot \bar{R}$ and $D_3 \cdot \bar{R}$, where the factors D_4 and D_3 are found in the table for sample size 5:

$$UCL = D_4 \cdot \bar{R} = (2.11)(16.7) = 35.2, \quad LCL = D_3 \cdot \bar{R} = (0)(16.7) = 0$$

The range chart below shows the range values in control:

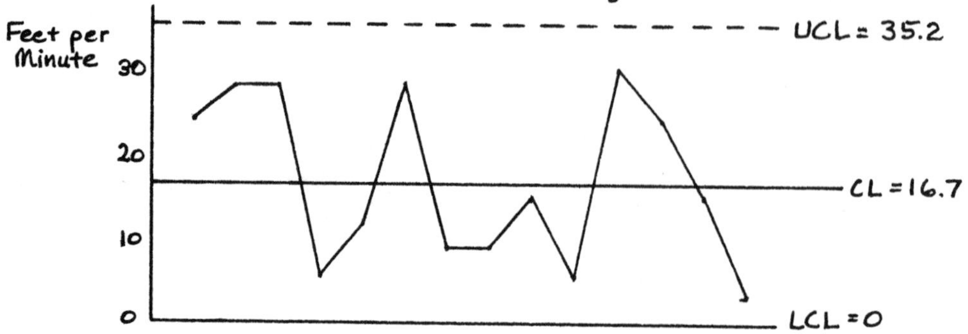

14-9

A process is known to have been in control with $\bar{x}' = 15$ and $\sigma' = 3$. Samples of size 9 will be taken to test for continued control. a) Find centerline and limits for both the $\bar{x}$ and R charts using 3σ limits. b) Find limits for the $\bar{x}$ chart using .99 and .01 probability limits.

****

a) $\bar{X}$ chart : $\sigma'_{\bar{x}} = \sigma'/\sqrt{n} = 3/3 = 1$; $3\sigma'_{\bar{x}} = 3$

therefore ℄ $= \bar{x}' = 15$; UCL $= 15 + 3 = 18$; LCL $= 15 - 3 = 12$

R chart: UCL $= D_2\sigma' = (5.39)(3) = 16.18$

℄ $= d_2\sigma' = (2.97)(3) = 8.91$

LCL $= D_1\sigma' = (.55)(3) = 1.65$

b) $Z_{.01} = 2.33$ so limits are $15 \pm 2.33\left(3/\sqrt{9}\right)$ < 17.33, 12.67

14-10

Use the data below to construct a fraction defective (p) control chart.

run	number produced	number defective
1	124	1
2	124	3
3	124	0
4	124	2
5	120	0
6	110	2
	$\Sigma = 726$	$\Sigma = 8$

****

$\bar{n} = \frac{726}{6} = 121$

℄ $= 8/726 = .01102 = p$

$$UCL = ℄ + 3\sqrt{\frac{p(1-p)}{n}} = .01102 + 3\sqrt{\frac{.01102(.98898)}{121}} = .0395$$

$$LCL = .01102 - 3\sqrt{\frac{p(1-p)}{n}} = -.0175 \Rightarrow 0$$

14-11 CONTROL CHARTS: FRACTION DEFECTIVE, p

The following numbers of defective valve seats were observed in 100% inspection of 12 days of production:

Number Inspected	Number Defective
148	5
105	5
172	9
86	6
125	8
140	7
74	5
186	10
110	7
98	5
133	3
123	2

Construct a stabilized p-chart for the data to determine if the process is in control.

****

Let p represent a sample fraction defective, $\bar{p}$ the average fraction defective over all samples, n the sample size (which varies). A point plotted on a stabilized p-chart is given by

$$\frac{p-\bar{p}}{\sqrt{\frac{\bar{p}(1-\bar{p})}{n}}}$$

with central line at 0 (zero) and control limits +3 (upper) and -3 (lower).

Sample calculation for first point: $p = \frac{5}{148} = 0.0338$, $\bar{p} = \frac{\Sigma \text{ defectives}}{\Sigma \text{ no. inspected}} = \frac{72}{1{,}500} = 0.048$ (remains the same for each point calculation).

$$\text{Point} = \frac{0.0338 - 0.0480}{\sqrt{\frac{(0.0480)(1-0.0480)}{148}}} = -0.809$$

Chart:

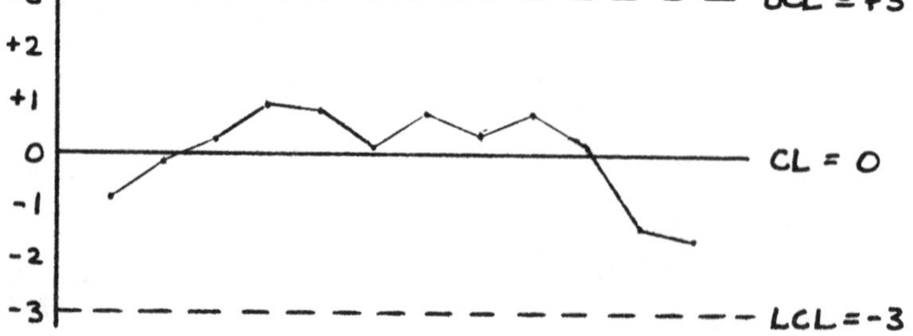

Analysis: The run of 8 points from Days 3-10 should be investigated for a shift upward in the fractions defective.

CONTROL CHARTS: NUMBER DEFECTIVE, *np*

14-12

Ten airplane wings were inspected for defective welds. The number of defects per wing range from 25 to 55. The total number of defects for the 10 units was 470.

Determine the average and the standard deviation of the Poisson distribution as required for the control limits of a c chart.

****

THE AVERAGE NUMBER OF DEFECTS PER WING

$$\bar{c} = \frac{470}{10} = 47$$

THE STANDARD DEVIATION OF THE POISSON DISTRIBUTION IS

$$\sigma = \sqrt{\bar{c}} = \sqrt{47} = 6.86$$

THE UCL AND LCL FOR THE C CHART ARE THE CUSTOMARY $\pm 3\sigma$ AWAY FROM $\bar{c}$, HENCE,

$$UCL = \bar{c} + 3\sqrt{\bar{c}}$$

$$\therefore \quad UCL = 47 + 3\sqrt{47} = 67.57$$

$$LCL = \bar{c} - 3\sqrt{\bar{c}}$$

$$\therefore \quad LCL = 47 - 3\sqrt{47} = 26.43$$

14-13

On the average about 3% of the bolts produced by a given automatic screw machine are defective. To maintain this quality of performance a sample of 100 bolts produced by the machine is examined every 2 hours.

Determine:

(a) The 99.73 control limits for the number of bolts that are defective in each sample.

(b) The 95% control limits for the number of bolts that are defective in each sample.

****

(a) $\bar{p} = 0.03, \quad n\bar{p} = 100(0.03) = 3$

$$\sigma = \sqrt{n\bar{p}(1-\bar{p})} = \sqrt{3(1-0.03)} = 1.706$$

$$UCL = n\bar{p} + 3\sigma = 3 + 3(1.706) = 8.118$$

$$LCL = n\bar{p} - 3\sigma = 3 - 3(1.706) \doteq 0$$

(b) $UCL = n\bar{p} + 1.96\sigma = 3 + 1.96(1.706) = 6.344$

$$LCL = n\bar{p} - 1.96\sigma = 3 - 1.96(1.706) \doteq 0$$

14-14

The following numbers of defective light bulbs were noted in 15 samples, each of size 50: 0, 3, 1, 1, 0, 2, 4, 1, 3, 5, 4, 3, 4, 5, 3. Determine if the process producing the light bulbs is in control according to an appropriate control chart.

****

An "np-chart" plotting numbers of defective light bulbs may be used. The central line (average) for the chart would be

$$n\bar{p} = \frac{0+3+1+1+0+2+4+1+3+5+4+3+4+5+3}{15} = 2.6$$

Control limits are given by $n\bar{p} \pm 3\sqrt{n\bar{p}(1-\bar{p})}$, where $\bar{p} = \frac{n\bar{p}}{50}$.

Upper control limit $= 2.6 + 3\sqrt{2.6\left(1-\frac{2.6}{50}\right)} = 7.3$

Lower control limit $= 2.6 - 3\sqrt{2.6\left(1-\frac{2.6}{50}\right)} = -2.1$ or 0

Control chart:

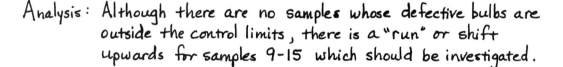

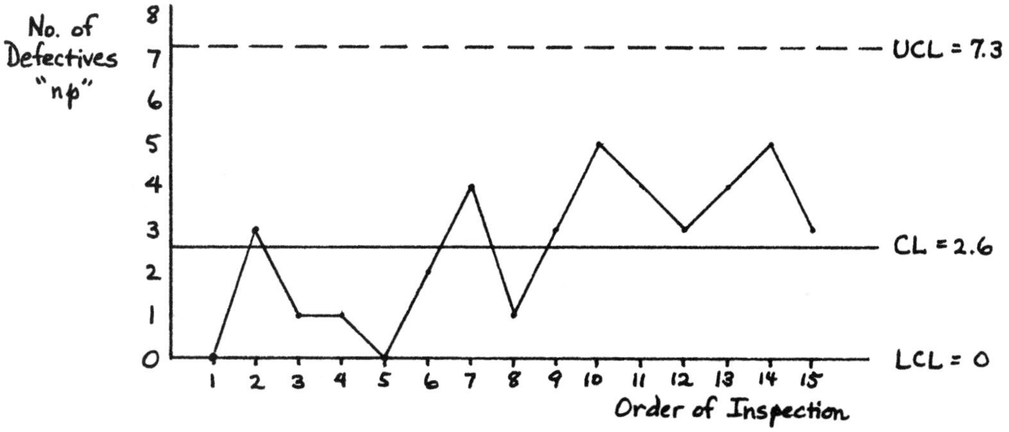

Analysis: Although there are no samples whose defective bulbs are outside the control limits, there is a "run" or shift upwards for samples 9-15 which should be investigated.

14-15

A manufacturer has designed a laser device to inspect a painting operation. The device will trip when the paint spray changes suddenly. The first ten units painted gave the data below. Construct the proper control chart and determine if the process is in control.

unit :	1	2	3	4	5	6	7	8	9	10
trips :	0	1	2	1	2	3	0	2	1	1

****

<u>np CHART</u>

$$\bar{c} = \frac{\Sigma c}{n} = \frac{13}{10} = 1.30$$

$$UCL = 1.30 + 3\sqrt{1.30} = 4.72$$

$$LCL = 1.30 - 3\sqrt{1.30} = -2.12 = 0$$

(ALL POINTS IN CONTROL)

ACCEPTANCE SAMPLING

14-16

A quality controller tests a random sample of 50 electric bulbs from a large batch, and accepts the batch if not more than two bulbs are defective. What is the probability of accepting a batch when

(i) 2% of the bulbs are defective
(ii) 20% of the bulbs are defective

****

Let p = probability of picking a bad bulb.

P(2 or less bad bulbs in a sample of 50) = P_a =

$$^{50}C_0 (1-p)^{50} + {}^{50}C_1\, p(1-p)^{49} + {}^{50}C_2\, p^2(1-p)^{48}$$

(i) $p = 0.02$

$$P_a = 0.98^{50} + 50 \times 0.02 \times 0.98^{49} + \frac{50 \times 49}{2} \times 0.02^2 \times 0.98^{48}$$

$$= \underline{\underline{0.922}}$$

(ii) $p = 0.2$

$$P_a = 0.8^{50} + 50 \times 0.2 \times 0.8^{49} + \frac{50 \times 49}{2} \times 0.2^2 \times 0.8^{48}$$

$$= \underline{\underline{0.001}}$$

14-17

A lot of expensive machine parts will be rejected by the purchaser if there is more than 5% defectives. The lot is sampled and the decision rule is to accept the entire lot if the sample proportion of defectives is less than 3%. How large a sample is required to be 99% sure of not accepting a bad lot? The null hypothesis should reflect the purchaser's reluctance to buy the lot.

****

$H_0: \pi \geq 0.05$ Lot is defective
$H_1: \pi < 0.05$ Lot is O.K.

where π is the (unknown) per cent defectives in the entire lot.

Decision Rule: Reject H_0 if the sample proportion $p < 0.03$

99% sure of not accepting bad lot implies a 1% probability or risk of rejecting H_0 when true,

i.e. $Pr(\text{Rej } H_0 | H_0 \text{ True}) = Pr(\text{Type I Error}) = \alpha = 0.01$

Under the Null hypothesis which can be considered as $H_0: \pi_0 = 0.05$ the sampling distribution of p is approximately Normal, i.e. $p \sim N(\mu_p, \sigma_p)$

where $\mu_p = \pi_0 = 0.05$ $\sigma_p = \sqrt{\dfrac{\pi_0(1-\pi_0)}{n}}$

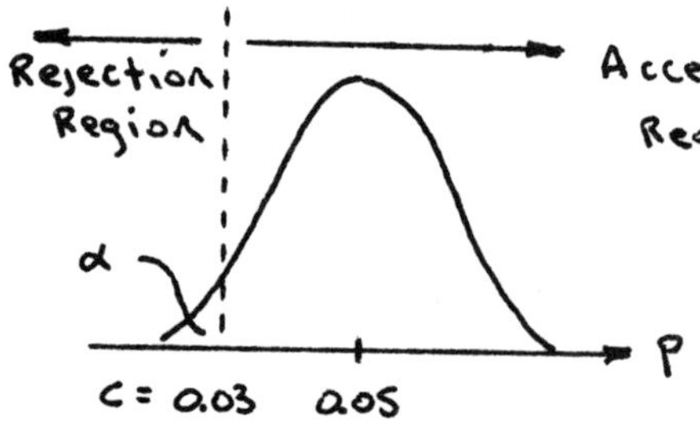

$$\sigma_p = \sqrt{\frac{(0.05)(0.95)}{n}}$$

$Pr(p \text{ is in rejection region} | \pi = 0.05) = 0.01$
$Pr(p < 0.03 | \pi = 0.05) = 0.01$

Standardizing p,

$$Pr\left[\frac{p-\mu_p}{\sigma_p} < \frac{0.03-0.05}{\sqrt{\frac{(0.05)(0.95)}{n}}}\right] = 0.01$$

$$Pr\left[Z < -0.02\sqrt{\frac{n}{0.0475}} \right] = 0.01$$

But $Pr(Z < z_{0.99}) = Pr(Z < -z_{.01}) = 0.01$

Hence, $-0.02\sqrt{\frac{n}{0.0475}} = -z_{0.01} = -2.33$

Solving for n, $n = 645$ to the nearest integer

Note, the large sample size justifies the use of the Normal distribution for the statistic p.

14-18

The double-sample acceptance sampling plan detailed below is used to inspect a particular purchased part which is received in very large lots. Determine the expected number of units inspected per lot from lots that are known to be 5% defective.

Sample number	Sample size	Acceptance number	Rejection number
1	32	0	3
2	48	3	4

****

Poisson approximation is appropriate.

Sample 1: $\lambda = np = (32)(0.05) = 1.6$

P(second sample required) = P(1 or 2 defectives on sample 1)

$= F(2) - F(0) = 0.783 - 0.202 = 0.581$

$ASN = 32 + (0.581)(48) = 59.9$

or $ASN = (32)(0.419) + (80)(0.581) = 59.9$

14-19

A manufacturing facility produces a shafts which are installed in a large gas turban. The mean diameter of the shafts produced is 400 mm. with a standard deviation of 10 mm. Although the standard deviation remains constant the diameter sometimes varies due to ware on the machine tool which produces the part. In order to detect such a shift in diameter the following quality control procedure has been established: Select a random sample of 36 shafts from the production lot and measure the diameter. If the diameter is more than 403 mm. or less than 397 mm. reject the production lot and make the appropriate adjustments to the tool .

a) What is the producers risk associated with the quality control plan?

b) If the process mean should shift to 395 mm. what is the probability that the shift will not be detected on the first sample after the shift?

****

a)

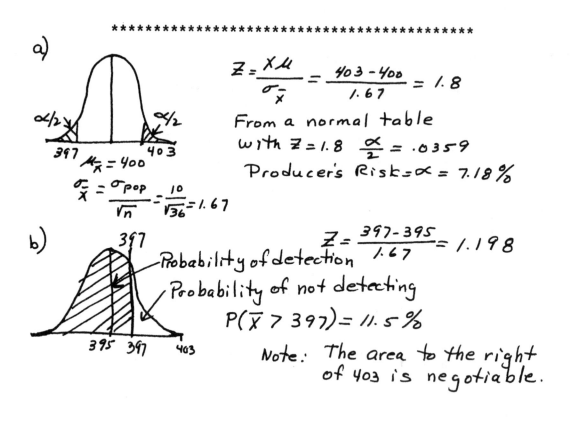

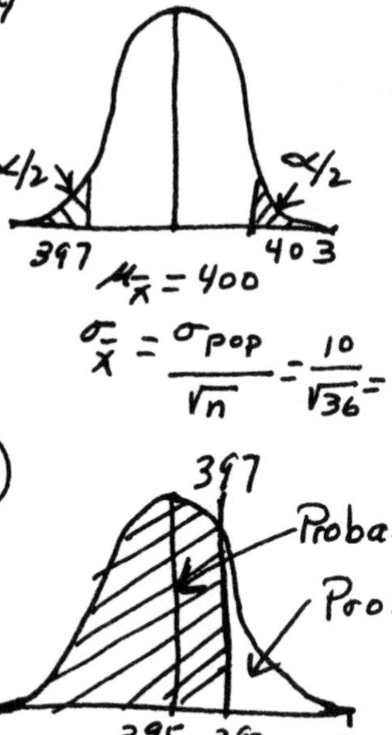

$\mu_{\bar{X}} = 400$

$\sigma_{\bar{X}} = \frac{\sigma_{pop}}{\sqrt{n}} = \frac{10}{\sqrt{36}} = 1.67$

$Z = \frac{X - \mu}{\sigma_{\bar{X}}} = \frac{403 - 400}{1.67} = 1.8$

From a normal table with $Z = 1.8$ $\frac{\alpha}{2} = .0359$

Producer's Risk $= \alpha = 7.18\%$

b)

Probability of detection

Probability of not detecting

$Z = \frac{397 - 395}{1.67} = 1.198$

$P(\bar{X} > 397) = 11.5\%$

Note: The area to the right of 403 is negotiable.

14-20

Estimate the average outgoing quality limit (AOQL) for a sample size of forty and an acceptance number of two. Assume a Poisson distribution.

****

$n = 40 \qquad c = 2$

FRACTION DEFECTIVE p	np	PROBABILITY OF ACCEPTANCE ** P_a	AOQ $P_a \cdot p$
.00	0	1.000	.0000
.02	.8	.953	.0191
.04	1.6	.783	.0313
.05	2.0	.677	.0339
.06	2.4	.570	.0342 ☆
.08	3.2	.380	.0304

** FROM POISSON TABLE

☆ AOQL ≃ .0342 @ $p = .06$

14-21

A double sampling plan is $n_1 = 50$, $c_1 = 0$; $n_2 = 60$, $c_2 = 3$, with rejection number 4. Determine the probability of acceptance of a 4% defective lot by use of a suitable approximation.

****

SCHEMATIC OF PLAN

1st	2nd
0	
1	2 OR LESS
2	1 OR LESS
3	0

Where $n_1 p' = 50(.04) = 2$

$n_2 p' = 60(.04) = 2.4$

USING THE POISSON APPROX.!

$P_A = .135 + (.271)(.570) + (.271)(.308) + (.180)(.091)$

$= 0.388$

14-22

Christmas tree bulbs are submitted in boxes of 200 for acceptance inspection under Military Standard 105D. Select a double sampling plan using the standard for Level II inspection with acceptable quality level (AQL) = 6.5%. Apply the plan to the following 35 consecutive boxes of bulbs, to determine which boxes would be rejected. Specify the plan used at each step of the acceptance sampling procedure.

Lot Number	1	2	3	4	5	6	7	8	9	10	11	12	13	14
No. of Defectives in 1st Sample	3	4	2	5	4	2	3	2	1	3	2	1	4	2
No. of Defectives in 2nd Sample	2	3	3	3	1	2	3	1	2	0	2	3	3	2

Lot Number	15	16	17	18	19	20	21	22	23	24	25	26	27	28
No. of Defectives in 1st Sample	3	4	2	3	4	0	1	0	1	3	0	1	1	0
No. of Defectives in 2nd Sample	2	2	3	5	1	2	1	1	3	0	3	2	4	1

Lot Number	29	30	31	32	33	34	35
No. of Defectives in 1st Sample	1	0	2	2	3	4	2
No. of Defectives in 2nd Sample	3	1	1	2	2	4	6

****

With AQL = 6.5%, Lot size 200, and Level II inspection, MIL-STD-105D begins with a normal double sampling plan having sample size 20 and acceptance (Ac) and rejection (Re) numbers on samples 1 and 2 of $Ac_1 = 2$, $Re_1 = 5$; $Ac_2 = 6$, $Re_2 = 7$. Under this plan, lot 2 is rejected on the second sample and lot 4 on the first sample. Lot 6 must therefore be inspected under a tightened plan with $Ac_1 = 1$, $Re_1 = 4$; $Ac_2 = 4$, $Re_2 = 5$. Lot 7 is rejected under this plan, then lots 8-12 are accepted, so a switch back to the normal plan is made beginning with lot 13. Lots 13 and 18 are then rejected, with lots 19-28 all accepted. Since 10 lots in a row are accepted, the total defectives in these 10 lots is compared to a table value in the standard (13 total exceeds 8 in table, so lot 29 is inspected under normal plan and accepted). Now the total defectives in lots 20-29 (the last 10) is 8 (equal to the table value), so a switch can be made beginning with lot 30 to a reduced plan (sample size = 8, $Ac_1 = 0$, $Re_1 = 4$; $Ac_2 = 3$, $Re_2 = 6$). Lots 30, 31, and 32 are accepted under the reduced plan; however, normal inspection is reinstated with lot 33 since the total defectives in lot 32's samples (4) fall between the Ac_2 and Re_2 numbers for the reduced plan. Under the normal plan, lot 34 is rejected.

Summarizing below, the plans under use are shown together with the sample (circled) at which the decision to accept (A) or reject (R) was made:

Lot No.	1	2	3	4	5	6	7	8	9	10	11	12	13	14	15	16	17	18
Sample 1 Def.	3	4	(2)	(5)	4	2	3	2	(1)	3	2	(1)	4	(2)	3	4	(2)	3
Sample 2 Def.	(2)	(3)	3	3	(1)	(2)	(3)	(1)	2	(0)	(2)	3	(3)	2	(2)	(2)	3	(5)
Decision	A	R	A	R	A	A	R	A	A	A	A	A	R	A	A	A	A	R

Lots 1–5: Begin with normal plan: $n = 20$, $Ac_1 = 2$ $Re_1 = 5$, $Ac_2 = 6$ $Re_2 = 7$

Lots 6–12: 2 of 5 lots rejected, so switch to tightened plan: $Ac_1 = 1$ $Re_1 = 4$, $Ac_2 = 4$ $Re_2 = 5$

Lots 13–18: 5 consecutive lots accepted—back to normal: $Ac_1 = 2$ $Re_1 = 5$, $Ac_2 = 6$ $Re_2 = 7$

Lot No.	19	20	21	22	23	24	25	26	27	28	29	30	31	32	33	34	35
Sample 1 Def.	4	(0)	(1)	(0)	(1)	3	(0)	(1)	(1)	(0)	(1)	(0)	2	2	3	4	(2)
Sample 2 Def.	(1)	2	1	1	3	(0)	3	2	4	1	3	1	(1)	(2)	(2)	(4)	6
Decision	A	A	A	A	A	A	A	A	A	A	A	A	A	A	A	R	A

Lots 19–29: Continue normal, but check total defectives after 10 consecutive accepted lots to see if a switch could be made to reduced inspection (this occurs after lot 29)

Lots 30–32: Reduced: $n = 8$, $Ac_1 = 0$ $Re_1 = 4$, $Ac_2 = 3$ $Re_2 = 6$

Lots 33–35: Back to normal since total defectives in samples 1 & 2 of lot 32 is between Ac_2 & Re_2 under tightened plan.

Rejected lots: 2, 4, 7, 13, 18, 34.

14-23

A certain kind of thread is to have a mean strength of 25.8 oz, σ being given as $\sigma = 2.34$. When tests are made with samples n = 16, a mean value of 25.0 oz. or below causes rejection of the lot. a) What is the size of the α risk? b) Compute the β risk for $\mu_1 = 24.5$ and $\mu_2 = 25.4$.

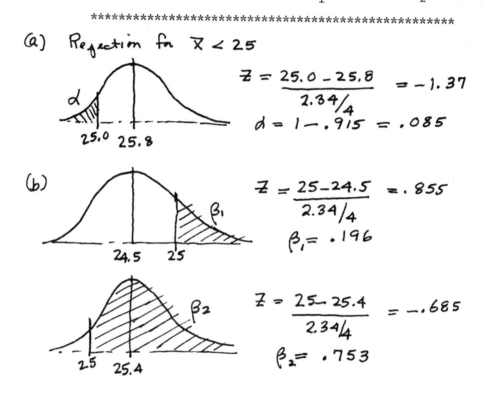

****

(a) Rejection for $\bar{X} < 25$

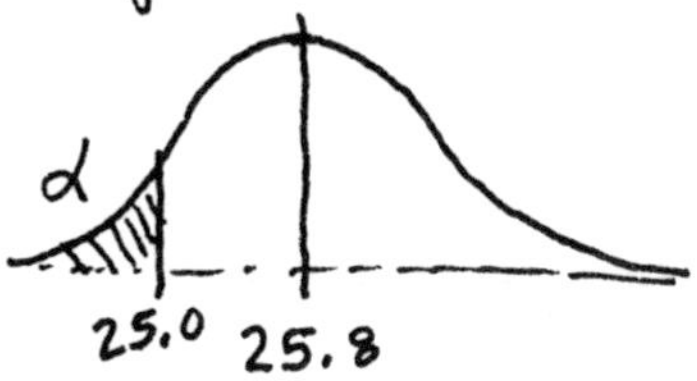

$$Z = \frac{25.0 - 25.8}{2.34/4} = -1.37$$

$$\alpha = 1 - .915 = .085$$

(b)

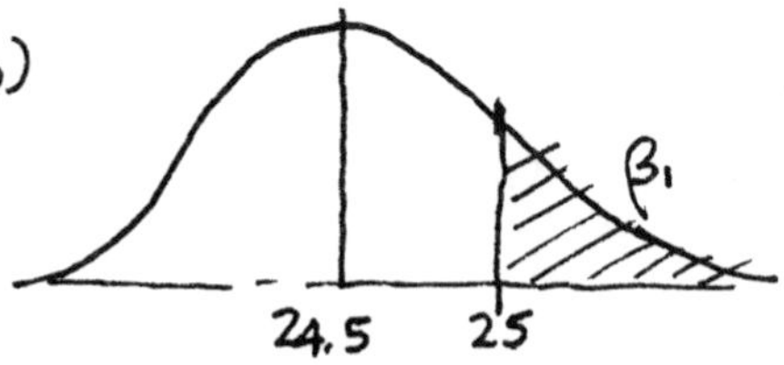

$$Z = \frac{25 - 24.5}{2.34/4} = .855$$

$$\beta_1 = .196$$

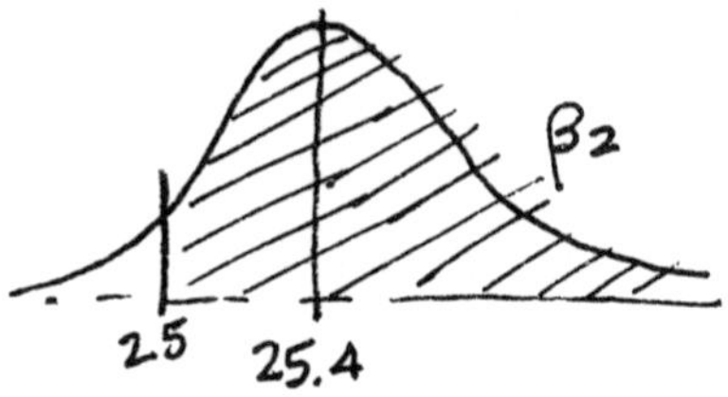

$$Z = \frac{25 - 25.4}{2.34/4} = -.685$$

$$\beta_2 = .753$$

15

RELIABILITY

SYSTEMS OF COMPONENTS WITH EXPONENTIAL FAILURE TIMES

15-1

Determine the reliabilities of the three simple systems shown: operating time for each is 200 hrs, unit failure rate is 2.5 x 10^{-4} failures/hour.

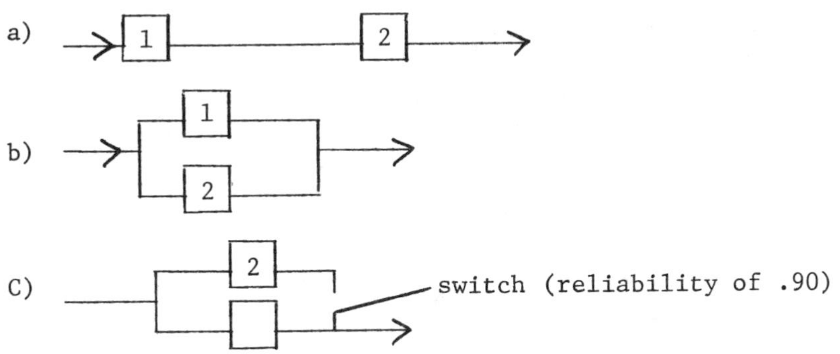

****

FOR EACH COMPONENT : $R(t) = e^{-\lambda t} = e^{-1/20} = .951$

(a) $\left(e^{-\lambda t}\right)^2 = .905$

(b) $1 - \left(1 - e^{-\lambda t}\right)^2 = 1 - .002 = .998$

(c) $e^{-\lambda t}\left(1 + \lambda t\, P_{SWITCH}\right) = .951\left[1 + (.05)(.90)\right] = .994$

15-2

A certain electrical component is considered reliable if it functions more than 50 hours. 24 such components were selected at random from the production line for testing. Their lives (in hours) prior to failure follow, recorded in the order of production: 55, 59, 51, 51, 53, 46, 57, 63, 54, 53, 49, 52, 59, 53, 56, 60, 48, 51, 54, 55, 51, 58, 52, 55. Estimate the reliability of a single component based on the sample, and then estimate the reliability of a subassembly consisting of three such components (each functioning independently).

****

Of the 24 components, 21 functioned beyond 50 hours.

Component reliability is thus estimated as $\frac{21}{24}$ = 0.875 or 87.5%

Subassembly reliability = $(0.875)^3$ = 0.67 or 67%

15-3

If a system must have reliability of 99%, how many components are required in parallel if each component has a reliability of 60 percent?

****

Unreliability of system = .01

" " each component = .40

$\therefore (.4)^m = .01$

$$m = \frac{\ln .01}{\ln .4} = 5.03 \rightarrow 6$$

15-4

Presume a three stage system with parallel/series redundancy and probabilities of successful operation for each assembly as shown below:

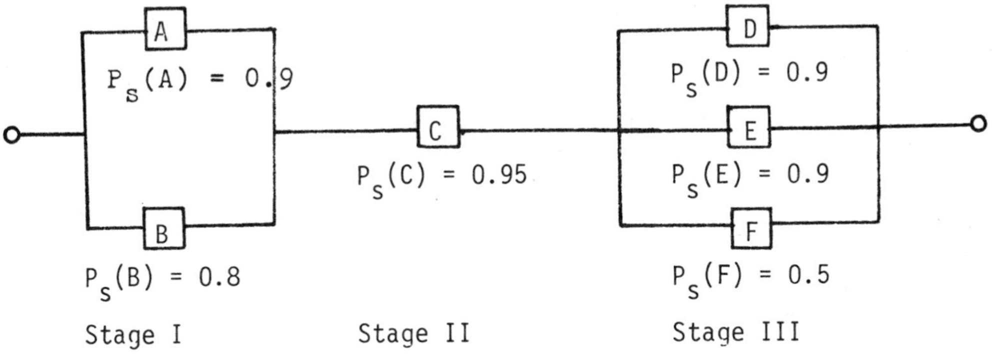

If the failure probabilities for each of the assemblies and each stage are independent, what is the system reliability?

****

$$R_{SYST.} = P\{(A \cup B) \cap C \cap (D \cup E \cup F)\}$$

$$= P(A \cup B) P(C) P(D \cup E \cup F)$$

$$= \{P(A) + P(B) - P(A \cap B)\} P(C) \{P(D) + P(E) + P(F) - P(D \cap E) - P(D \cap F) - P(E \cap F) + P(D \cap E \cap F)\}$$

$$= \{P(A) + P(B) - P(A)P(B)\} P(C) \{P(D) + P(E) + P(F) - P(D)P(E) - P(D)P(F) - P(E)P(F) + P(D)P(E)P(F)\}$$

$$\therefore R_{SYST.} = \{0.9 + 0.8 - 0.9(0.8)\} 0.95 \{0.9 + 0.9 + 0.5 - 0.9(0.9) - 0.9(0.5) - 0.9(0.5) + 0.9(0.9)0.5\}$$

$$= 0.98(0.95)0.995 = 0.926^{+}$$

15-5

An electronic device has a failure rate of 400 failures/10^6 hours. Two identical units form a system with one operating, the other on standby. The reliability of the system's failure-sensing and switching unit is $P_{FSS} = 0.98$.

What is the system's reliability as a function of t?

****

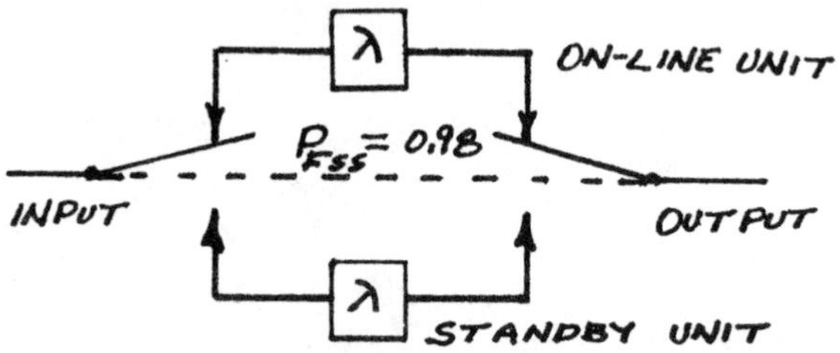

$$P_0(t+\Delta t) = P_0(t)(1-\lambda\Delta t) \Rightarrow \lim_{\Delta t \to 0} \frac{P_0(t+\Delta t)-P_0(t)}{\Delta t} = -\lambda P_0(t)$$

$\therefore \frac{dP_0(t)}{dt} = -\lambda P_0(t)$, FOR THE INITIAL CONDITIONS

$$P_0(t=0)=1, \text{ AND } P_1(t=0)=0, \; P_2(t=0)=0$$

$$\frac{dP_0(t)}{P_0(t)} = -\lambda dt \Rightarrow \ln P_0(t) = -\lambda t + C_0$$

$$\ln 1 = -\lambda \cdot 0 + C_0 \quad \therefore C_0 = 0, \; P_0(t) = e^{-\lambda t}$$

$$P_1(t+\Delta t) = P_1(t)(1-\lambda\Delta t) + P_0(t)\lambda\Delta t \Longrightarrow$$

$$\lim_{\Delta t \to 0} \frac{P_1(t+\Delta t) - P_1(t)}{\Delta t} = -\lambda P_1(t) + \lambda P_0(t)$$

$$\therefore \frac{dP_1(t)}{dt} = -\lambda P_1(t) + \lambda e^{-\lambda t}, \quad \frac{dP_1(t)}{dt} + \lambda P_1(t) = \lambda e^{-\lambda t}$$

USING THE INTEGRATING FACTOR $e^{\int \lambda dt} = e^{\lambda t}$

$$\therefore e^{\lambda t}\frac{dP_1(t)}{dt} + e^{\lambda t}\lambda P_1(t) = \lambda \Rightarrow \frac{d}{dt}\left[e^{\lambda t}P_1(t)\right] = \lambda$$

$$e^{\lambda t}P_1(t) = \lambda t + C_1, \text{ AT } t=0 \quad C_1 = 0$$

$$\therefore P_1(t) = \lambda t e^{-\lambda t}$$

$$R_{SYST}(t) = P_0(t) + P_{FSS} P_1(t)$$
$$= e^{-\lambda t} + P_{FSS} \lambda t e^{-\lambda t}$$
$$= e^{-0.0004t} + 0.98(0.0004t)e^{-0.0004t}$$
$$= (1 + 0.000392t)e^{-0.0004t}$$

15-6

Suppose that a particular electronic system consists of four components (A, B, C and D) connected in parallel. The time to failure for each component is exponentially distributed with mean times between failures of 200, 200, 300 and 600 hours, respectively. Determine the system reliability for 1000 hours, assuming that each component operates independently and that each begins operation when the system begins operation.

****

Component	P(failure in 1000 hours)
A, B	$1 - e^{-1000/200} = 0.993$
C	$1 - e^{-1000/300} = 0.964$
D	$1 - e^{-1000/600} = 0.811$

System reliability = 1 − P(all components fail)

$$= 1 - [(0.993)^2(0.964)(0.811)] = 0.229$$

15-7

The time to failure of a machine part is known to obey exponential probability distribution. If the probability that the part would not survive for more than 30 days is 0.9, how often is the machine part expected to be replaced?

Note: This knowledge is useful in spare-part inventory planning for machinery maintenance.

****

If time to failure T obeys exponential distribution, the cumulative distribution function is given by

$$P(T \leq t) = F(t) = 1 - e^{-\lambda t} \quad \text{for } t \geq 0$$
$$= 0 \quad \text{for } t < 0$$

The mean $E(T) = 1/\lambda$

Its is given that

$$P(T \leq 30) = 0.9$$

$$\therefore \quad 0.9 = 1 - e^{-\lambda \times 30}$$

Hence $\lambda = -\frac{1}{30} \ln(1-0.9) = 0.0768$

$$\therefore \quad E(T) = \frac{1}{\lambda} = \underline{\underline{13.0 \text{ days}}}$$

The machine part is expected to be replaced every 13 days.

15-8

Components A and B have exponentially distributed lifetimes with a mean of 30 hours and 45 hours, respectively. In order to perform the same operating function 10 identical components of type A must be in series whereas 15 identical components of type B are needed. Which design is better?

****

$\mu_A = 30$; $\mu_B = 45$; $T_A \sim exp(1/30)$; $T_B \sim exp(1/45)$

S_A = LIFETIME OF 10 COMPONENTS OF A IN SERIES

S_B = LIFETIME OF 15 COMPONENTS OF B IN SERIES

$$R_{S_A}(t) = P\{S_A > t\} = \left[e^{-t/30}\right]^{10} = e^{-t/3}$$

$$R_{S_B}(t) = P\{S_B > t\} = \left[e^{-t/45}\right]^{15} = e^{-t/3}$$

PROBABILISTICALLY, THE DESIGNS ARE IDENTICAL.

15-9

In a room containing six computer terminals (each with a mean time to failure of eight months), what is the chance of two or fewer failures in a four month period?

****

POISSON DISTRIBUTION

$$E(\text{NUMBER OF FAILURES}) = 6\left(\tfrac{1}{8}\right)4 = 3$$

$$P(\text{2 OR FEWER FAILURES}) = \sum_{i=0}^{2} \frac{e^{-3}\,3^i}{i!}$$

$$= e^{-3}\left[\frac{3^0}{1} + \frac{3^1}{1} + \frac{3^2}{2}\right] = 42.32\ \%$$

15-10

An electronic device is made up of four components, two of which are connected in parallel which in turn are connected in series with the other two components. This arrangement is schematically shown in the figure.

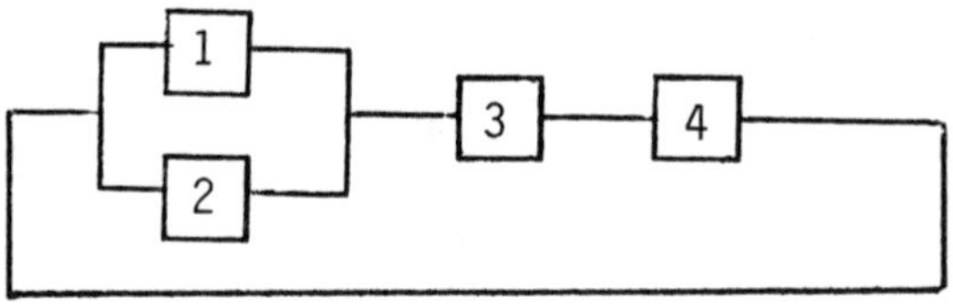

Reliabilities of the four components are denoted by $R_1(t)$, $R_2(t)$, $R_3(t)$, and $R_4(t)$. System failure occurs either when both parallel components fail or when any one of the two series components fail. Assuming that the component failure takes place independently, give a suitable reliability model for this device.

If the probability of failure in each component can be represented by the exponential distribution

$$F(t) = 1 - e^{-\lambda t} \qquad \text{for } t > 0$$

having the same failure rate $\lambda = 0.01/\text{hr}$, what is the probability that the device would not survive for more than 100 hrs?

****

Reliability $R(t) = 1 - F(t)$

Where

$F(t)$ = Probability of failure at or before time t

For series components;

overall reliability = product of component reliabilites

For the parallel component;

overall prob. of failure = product of component prob. of failure

It follows that for the parallel unit consisting of 1 and 2

Prob. of failure $= (1-R_1)(1-R_2)$

$\therefore$ its reliability $= 1-(1-R_1)(1-R_2)$

Since this parallel unit is connected in series with the components 3 and 4, the overall reliability of the device is

$$R(t) = [1-(1-R_1)(1-R_2)]R_3 R_4$$

For the exponential model

$$F_1(t) = F_2(t) = F_3(t) = F_4(t) = 1 - e^{-0.01t}$$

Hence

$$R_1(t) = R_2(t) = R_3(t) = R_4(t) = e^{-0.01t}$$

$$\therefore R(t) = [1-(1-e^{-0.01t})(1-e^{-0.01t})]\, e^{-0.01t} e^{-0.01t}$$

$$= (2e^{-0.01t} - e^{-0.02t})e^{-0.02t} = (2-e^{-0.01t})e^{-0.03t}$$

The prob. of failure

$$F(t) = 1 - R(t) = 1-(2-e^{-0.01t})e^{-0.03t}$$

With $t = 100$ hrs

$$F(100) = 1-(2-e^{-1})e^{-3}$$

$$= 0.92$$

15-11

What is the probability that a light bulb will survive at least double its expected life?

****

EXPONENTIAL DISTRIBUTION

$$P(x > 2u) = e^{-1/u\,(2u)} = e^{-2} = 13.53\%$$

LIFE TESTING: EXPONENTIAL MODEL

15-12

Fifteen identical microprocessor chips were tested simultaneously under the same severe stress environment for 48 hrs and the time to failure (in hours) for each unit that failed was recorded, as given below:

10.2, 15.6, 20.0, 32.5, 36.8, 36.9, 37.1,
37.8, 40.1, 42.5, 45.3, 47.5, 47.9

Note that two units did not fail during this period. Estimate the probability that a microprocessor chip of this type would survive for 36 hrs when subjected to a similar high-stress environment.

****

For a conservative estimate use 50 hrs as the failure time of the two survived chips. Hence the average time to failure

$$\bar{t} = \frac{1}{15}[10.2 + 15.6 + \cdots + 47.9 + 2 \times 50] = 36.68 \text{ hr}$$

We shall employ the exponential model with failure rate λ. The mean time to failure

$$\int_0^\infty t \lambda e^{-\lambda t} dt = \frac{1}{\lambda} \simeq 36.68$$

$$\therefore \quad \lambda = \frac{1}{36.68} \quad \text{(as an estimate)}$$

Prob. of failure in t hrs or less is

$$F(t) = 1 - e^{-\lambda t} = 1 - e^{-\frac{t}{36.68}}$$

$\therefore$ Prob. of survival for at least t hrs is

$$1 - F(t) = R(t) = e^{-t/36.68}$$

Now with $t = 36$ hrs.

$$R(36) = e^{-36/36.68} = \underline{\underline{0.375}}$$

LIFE TESTING: WIEBULL MODEL

15-13

Two designs for a critical component are being studied for adoption. From extensive testing of prototypes, it was found that the time to failure follows a two parameter Weibull distribution. Design A costs $4000 to build and has Weibull shape and scale parameters of 3 and 190 hours, respectively. Design B costs $3000 to build and has Weibull shape and scale parameters of 4 and 100 hours, respectively. Which design should the manufacturer produce if the component must have (a) a 10 hour guaranteed life, (b) a 20 hour guaranteed life?

****

$$F_A(t) = 1 - e^{-(t/190)^3} \;; \qquad F_B(t) = 1 - e^{-(t/100)^4}$$

$$R_A(10) = e^{-(10/190)^3} = .99985 \;; \quad R_B(10) = e^{-(10/100)^4} = .99990$$

a) B IS SLIGHTLY MORE RELIABLE AND COST LESS ALSO.

$$R_A(20) = e^{-(20/190)^3} = .99883 \;; \quad R_B(20) = e^{-(20/100)^4} = .99840$$

b) WHILE B IS SLIGHTLY LESS RELIABLE, THE COST WOULD PROBABLY CAUSE B TO BE FAVORED OVER A.

15-14

An automaker wanted to conduct a statistical study on in-service cracks developed in some welded joints in the chasis of a particular model of pickup truck under normal operation. The company selected a representative sample of 100 trucks for the test. The trucks were called in every 6 months and inspected to detect weldment cracks, and the cumulative number of trucks that developed cracks was recorded. Test data gathered over a period of 3 years are given below:

Months Operated (t)	6	12	18	24	30	36
No. of Trucks that Developed Cracks (n)	4	17	35	55	72	84

Using a suitable failure model, estimate the probability that the cracks would develop during the warranty period of 1 year.

****

Weibull failure model is appropriate here because the cracks are quite likely due to fatigue. Assume that there is no infant mortality failure.

Probability of failure within time t ;

$$F(t) = 1 - e^{-\lambda t^s}$$

where, λ = Weibull scale parameter
s = Weibull shape parameter

Reliability $R(t) = 1 - F(t) = e^{-\lambda t^s}$

$$\therefore \quad \ln R(t) = -\lambda t^s$$

or

$$\ln[-\ln R(t)] = s \ln t + \ln \lambda$$

Note that the curve $\ln[-\ln R(t)]$ vs. $\ln t$ is a straight line with slope s and y-intercept $\ln \lambda$ for this model. Hence s and λ could be experimentally determined using such a plot.

If m trucks are tested and n are found to develop cracks in time t or less, an estimate for the probability of failure within time t would be

$$\hat{F} = \frac{n}{m}$$

An estimator for reliability would be

$$\hat{R} = 1 - \hat{F} = 1 - \frac{n}{m} = \frac{m-n}{m}$$

For the given data m = 100. The numerical values for $\ln[-\ln \hat{R}]$ and $\ln t$ are computed and tabulated below:

t	n	$\hat{R}$	$\ln(-\ln \hat{R})$	$\ln t$
6	4	0.96	−3.20	1.79
12	17	0.83	−1.68	2.48
18	35	0.65	−0.84	2.89
24	55	0.45	−0.23	3.18
30	72	0.28	0.24	3.40
36	84	0.16	0.61	3.58

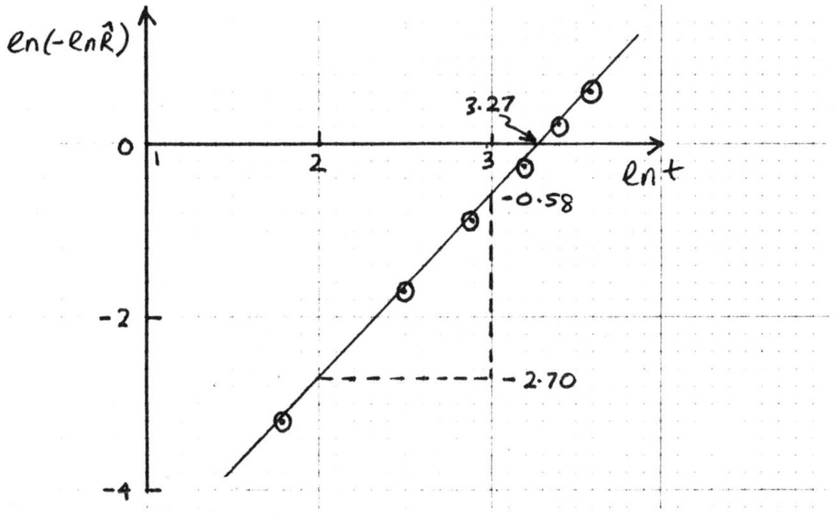

The values are plotted and a straight-line fit is drawn as shown.

Slope $s = -0.58 - (-2.70) = 2.12$

By substituting the intercept on the x axis into the straight line equation;

$$0 = 2.12 \times 3.27 + \ln A$$

Hence

$$A = 0.001$$

Thus the estimated model is given by

$$F(t) = 1 - e^{-0.001t^{2.12}}$$

Probability of failure within one year ($t = 12$) is

$$F(12) = 1 - e^{-0.001 \times 12^{2.12}}$$

$$= \underline{\underline{0.18}}$$